WINNING
THE GAMES
SCIENTISTS PLAY

Other Trade Books by Carl J. Sindermann

The Scientist as Consultant (with Thomas K. Sawyer)
The Woman Scientist (with Clarice M. Yentsch)
The Joy of Science
Survival Strategies for New Scientists

WINNING THE GAMES SCIENTISTS PLAY

Strategies for Enhancing Your Career in Science

Carl J. Sindermann, Ph.D.

PERSEUS PUBLISHING • Cambridge, Massachusetts

Many of the designations used by manufacturers and sellers to distinguish their products are claimed as trademarks. Where those designations appear in this book and Perseus Books was aware of a trademark claim, the designations have been printed in initial capital letters.

A CIP record for this book is available from the Library of Congress.

Copyright © 2001 by Carl J. Sindermann

ISBN: 0-7382-0425-0

Perseus Publishing is a member of the Perseus Books Group.

Find us on the World Wide Web at http://www.perseuspublishing.com

Perseus Publishing books are available at special discounts for bulk purchases in the U.S. by corporations, institutions, and other organizations. For more information, please contact the Special Markets Department at the Perseus Books Group, 11 Cambridge Center, Cambridge, MA 02142, or call (617)252-5298.

Text design by Bookcomp, Inc.
Set in 10-point Palatino by Bookcomp, Inc.

1 2 3 4 5 6 7 8 9 10—03 02 01
First printing, March 2001

I offer this latest version of my
favorite book to my wife Joan, who has
always been a source of encouragement for
my periodic literary outbursts.

PREFACE
To The Revised Edition

The interpersonal strategies that surround the act of doing good science—hereafter referred to loosely as "scientific game playing"—have received sporadic published attention, and many game rules have become almost axiomatic among successful practitioners of science. There is a need, however, to review periodically what we know and what we think we know about the art, and to add new insights as they become available. This new edition of *Winning the Games Scientists Play* is a response to that need; it has been written for science practitioners and onlookers of the new century, drawing on insights and perceptions gained from victories and defeats of the waning decades of the 20th century.

Let me make it perfectly clear at the outset that this book is *not* about games with rules that can be formalized mathematically. It has nothing to do with "game theory" as developed by mathematicians at Princeton and elsewhere beginning with the work of von Neumann in 1944. Its strategies cannot be analyzed quantitatively, because they are essentially subjective, and constitute in their best expression an art form rather than a formal discipline.

It seems especially important that the strategies and rules of scientific game playing be reviewed critically as we move into the next millennium, since some (but not all) of those rules have undergone varying degrees of change during the 1980s and 1990s. Those decades saw shrinkage in faculty size and in federal grants. The university climate has been less than perfect. Long lines of candidates have formed for good and even mediocre faculty positions; good Ph.D.'s languish interminably in post-doc positions; tenure has been questioned; and part-time instructors have proliferated amazingly on many campuses.

In times of stringency scientists adapt, and scientific game playing achieves new heights of precision and worth. This book is an introduction to the new era, borrowing from the relevant past, for scientists old and new who will be (as politicians and bureaucrats are fond of saying) "facing the challenge of the future." It is an attempt to blend some serious, hard-earned, occasionally misconceived, but usually practical advice on selected topics important to a professional career, with other, less serious, and probably equally misconceived principles of interpersonal behavior in science. The goal, in other words, is to achieve a readable mix of the "how to" with the "joy of" science—the techniques with the pleasures, and the realities with the fantasies. At times it may seem difficult or downright impossible to separate fact from fiction in some of the following chapters; I suggest that in those instances the choice should be fiction, since there is little that is factual (in the sense of being amenable to conclusive demonstration or quantification) in this book. Treat it like a book of poems, some of which may have meaning and relevance to you, and some of which are merely words pinned to a line in a maze. Much of what is described herein could probably be defined as "common sense" but it seems not to be "common" in the sense of universal understanding or acceptance of the pertinent rules of scientific games presented.

Some may feel after reading this book that the real world of science is simpler and purer than that portrayed here—that there is less opportunism, fewer calculated steps, and none of the maneuvering described in this book. I think this is a naive (one of the worst epithets in science) viewpoint, but I would deny no one his or her comfort blanket. I simply want to make explanations available when the blanket is snatched away, as it almost inevitably will be, usually by a practicing player. Whether a scientist chooses to participate in games or not, he or she should be aware of their nature, if only to avoid being drawn in unwittingly or being used as a pawn in someone else's game.

It should, however, be absolutely clear that I am in no way and at no time denigrating honest, persistent scientific effort, sustained research productivity, or brilliant insights, since these constitute the real substance of science. I am simply outlining another dimension which, if taken into account, can enhance those fundamental characteristics and make the joys of a scientific career even greater. Stated succinctly, *the core of game playing is starting with a base of scientific excellence and making the most of it by following simple rules.* Game players must have credibility and integrity to be successful—otherwise they will be looked upon as manipulators and "operators."

Writing this new edition of a book on the interpersonal games that scientists play has provided me with a unique opportunity to restructure the universe to fit my own personal perceptions—and to add a few insights as well as to make a minimal attempt to correct some outmoded attitudes (my wife refers to them as "anachronistic" attitudes). Many scientists have contributed, wittingly or unwittingly, to this effort, but they will remain forever anonymous (often at their own strongly worded insistence) although they should recognize themselves somewhere in the text.

I would like to thank Perseus Books for offering me a rare occasion to reassess all those firmly held convictions presented so confidently in the first edition, and to tweak a few elements to make the present edition a good source book for a world of science and scientists that has just plunged into a new century.

I would also like to acknowledge the hospitality of the

Commonwealth of Massachusetts, for providing facilities for writing and contemplation at South Pond in the Savoy Mountain State Forest, high in the northern Berkshires. Without drawing too many gratuitous parallels, South Pond is in many of its characteristics the present-day equivalent of the well-known, but now despoiled, Walden Pond (located in the eastern part of the Commonwealth) as it was more than a century ago during Henry David Thoreau's tenancy there.

The development of this new edition has been a source of great pleasure and stimulation for me, and has led to the appearance of a few new perspectives on the interpersonal areas of science as practiced in this country and elsewhere—as well as to a comforting reaffirmation of some strongly held earlier viewpoints.

Carl J. Sindermann

Easton, Maryland
September 1, 2000

CONTENTS

Introduction: The Importance of Interpersonal Strategies
in Science   1

PART ONE: A PRIMER FOR SCIENTIFIC
STRATEGISTS   25

Chapter 1: The Scientist as a Writer: Publishing
Scientific Papers   27

Authorship of scientific papers   28
The role of reviewers   40
Scientific writing viewed from the editor's
desk   45
Damming the paper flood   49
Summary   52

Chapter 2: The Scientist as a Performer:
Presenting Scientific Papers   53

How not to present a scientific paper   54
Suggestions for better paper presentations   56
How to enjoy presenting a scientific paper   64
The so-called "discussion" following paper
presentation   67
The last word   69

Chapter 3: The Scientist as a Face in the Crowd: Attending Scientific Meetings   71

    Strategies for meeting participation   71
    Funding meeting attendance   78
    Election to society offices   79
    International meetings   81
    Residues of scientific meetings   86
    Summary   88

Chapter 4: The Scientist as a Concertmaster: Chairing Scientific Sessions   89

    Early planning   91
    The meeting   94
    Session follow-up   98
    Special assignments   100
    Summary   104

Chapter 5: The Scientist as a Producer/Director: Organizing Scientific Meetings   105

    The annual meeting of a scientific society   106
    The workshop or small-group conference   113
    The international symposium   117
    Summary   123

Chapter 6: The Scientist as a Negotiator: Participating in Committee Meetings   125

    The chairperson's creed   126
    The committee members' guide   128
    Some committee games   130
    Committee women   132
    The advisory committee   133
    Paper products of committee
       activities   135
    Computer conferences   137
    Summary   140

PART TWO: CRITICAL ISSUES FOR SCIENTIFIC STRATEGISTS   141

Chapter 7: The Scientist in Transition: Moving Up, On, and Out   143

The fast track   146
Leaving the fast track   148
Life stages of a scientist   150
Eras in a professional career   155
The "professional"   158
Losing strategies   163
Moving on   166
Moving out   170
Summary   172

Chapter 8: The Scientist in Control: Getting and Using Power   175

Kinds of power in science   178
Power strategies   183
Organizational variations in power   187
Summary   189

Chapter 9: The Scientist in Doubt: Defining Ethics in Science   191

The perimeter of the circle   192
Within the circle   197
Maneuvers in ethical combat zones   199
Summary   206

PART THREE: SPECIAL INTEREST AREAS FOR SCIENTIFIC STRATEGISTS   207

Chapter 10: Women in Science: A Current Appraisal   209

The movement toward equality: current status of women in science   210
Impediments to achieving full equality for women scientists   215
Women scientists in the corridors of power   218
Summary   222

Chapter 11: Coping with Bureaucracy and
    Bureaucrats   225

    Types of bureaucrats   226
    Common characteristics of bureaucrats   234
    The bureaucratic hierarchy   237
    Kinds of bureaucracies   239
    Summary   247

Chapter 12: Dealing with External Forces: News Media,
    Lawyers, Politicians, and the Public   249

    News Media   250
    Lawyers   253
    Politicians   255
    The public   257
    Summary   259

Chapter 13: The Scientist in Industry   261

    Industrial game rules   263
    Kinds of industrial research laboratories   264
    Consulting   269
    The scientist as an industry spokesperson   271
    Summary   273

Conclusions   275

References   279

Index   283

THE IMPORTANCE OF INTERPERSONAL STRATEGIES IN SCIENCE

This book has been written for thousands of serious, dedicated scientists—particularly the new ones, but including those at any stage of development—to ensure that they participate fully and joyously in a genuinely wonderful occupation, which, like many others, operates best within the confines of a set of rules. It has been written also for the spouses, families, and friends of scientists, so that all may understand the larger environment in which science is embedded.

"Game playing," in science or in any occupation, can be defined as the *ability and willingness not only to govern your actions by a set of rules, but to have the rules work in your favor.* It is a legitimate approach to a satisfying life in science, based on the assumption that you have intelligence, energy, training, and perception—the normal basic equipment for a scientific career. Alternatively, games may be characterized as "interpersonal strategies" or "prescribed procedures for standard situations" or "codified responses to actions by others" or "competitions conducted according to rules," but whatever the definition, games that surround the act of doing good science do exist, and can be identified and enjoyed.

It is logical to learn—even to memorize—the rules of a

game if you are to be a serious participant. Beyond this you must have ability, insight, and ingenuity if you are to be a professional player. *Many scientists play like amateurs or like poorly paid semipros throughout their entire careers because they do not pay adequate attention to simple game rules—or worse still, because they deny the existence of rules, or the presence of any formalized structure for interpersonal and institutional relations in science.*

Most scientists follow a prescribed developmental pattern: graduate school, orals, dissertation, dissertation defense, degree, then—poof—they are professionals, but without formal courses or seminars in the very critical area of professional game playing, despite the fact that their major professors may be expert players. Such faculty members expect their students to learn by example, which the more observant ones are able to do but which others fail to do.

I should point out clearly that there are many viewpoints about the proper role of game playing in science—but there should be no denial that it does play a role. The contents of this book represent an attempt to codify a fund of observations gained as a spectator and participant on some of the great playing fields of science—cocktail parties, symposium mixers, review committee meetings, laboratory conferences, seminars, workshops, faculty committee meetings, international symposia, and many others.

Game activities are not always overt, and the rules are freqently imprecise or arguable; sometimes the real pleasure is in *playing*, even though winning is the normal objective. Creative people—scientists included—enjoy what they do best; the "highs" often come in manipulations of concepts, systems, or analyses. These same people also enjoy participating in the interpersonal games described in these chapters.

So let's begin this book with a few premises which you may conclude are false before you reach page 50: (1) game rules exist; (2) a broad spectrum of variation also exists; and (3) denial of the existence of rules is perilous. Let's begin too with the axiom that *a scientific career is fundamentally pleasurable and satisfying, but can be infinitely more rewarding with the proper application and acceptance of game rules by all participants.*

To me, *a game* (as exemplified in science) *constitutes a series of transactions and strategies which legitimately enhances progress in the many interpersonal relations which surround the act of doing good science.*

People interactions, then, are important in doing science; relationships with peers, colleagues, supervisors, assistants, secretaries, and a host of others help determine our success or failure. We bring to the science workplace all (or at least most) of our good and bad personal characteristics; we do not metamorphose into something different or purer. We may in some instances be on good behavior, temporarily suppressing evil tendencies in the interest of getting on with research production. Usually, though, long-term close contact in the classroom or laboratory, as in any work environment, leads to a high degree of self-exposure. Periods of stress are particularly effective in the disclosure process, sometimes revealing astounding facets of our thinking or character.

The objective of this book is to explore, with some reasonable good humor, balance, and insight, the complex subject of interpersonal relationships in science. It is an attempt to summarize the non-technical career strategies of science professionals, and to offer remote counseling for those interested in winning the games scientists play.

The volume is divided into three principal sections: Part One, "A Primer for Scientific Strategists"; Part Two, "Critical Issues for Scientific Strategists"; and Part Three, "Special Interest Areas for Scientific Strategists." The "Primer" section, predictably, includes chapters on writing and publishing scientific papers, presenting scientific papers, attending scientific meetings, chairing scientific sessions, organizing scientific meetings, and participating in committee meetings. The "Critical Issues" section includes chapters on moving up, on, and out; getting and using power; and defining ethics in science. The "Special Interest" section includes a completely new chapter on women in science; as well as chapters on coping with bureaucracy and bureaucrats; dealing with the news media, lawyers, politicians, and the public; and scientists in industry.

In examining this substantial core of information about personal relationships of scientists, it has become apparent to me that recent decades have produced a number of changes in how science professionals interact with each other and with the world, and that these should be identified early in this discussion.

One clear example is the *evolving role of women in science* during the past three decades (1970–2000). It's an emerging success story that will get greater attention in Chapter 10—still a work in progress—but movement toward equality has been substantial. Other changes in how scientists think and act have moved onstage during the decades of the 1980s and 1990s. During that period we witnessed the *great expansion of consulting* as a significant career option for science professionals; and we also recognized the ever-increasing need for multidisciplinary team research, with consequent demands on lead scientists to become *professional managers* as well. Then, during the decade of the 1990s, we have seen the explosive growth of the Internet and other forms of *electronic communication* in globalized technical information transfer; and many of us have even felt a need to participate in the burgeoning development of *biotechnology*, which is invading many areas of scientific investigation, and has truly awesome potentials.

Each of these relatively recent shifts in the rhythm of the dance has had ripple effects on the ways scientists perceive, communicate with, and relate to each other. Women are no longer considered only as technicians or excluded from high professional ranks; a number of scientists have demonstrated entrepreneurial skills that have made them rich as well as famous; science management—especially of "big science" programs—has proved to be an extreme test of organizational competence; the Internet has transformed access to technical and all other forms of information; and biotechnology with all its ramifications promises to make fundamental alterations in many of the subdisciplines of science. These changes in how science is conducted should not, however, be allowed to obscure the truism that *the rules of behavior that most members of the scientific community will accept are essentially conservative, even though the practitioners are engaged in a highly dynamic and innovative enterprise.*

As a brief introduction to a way of thinking about science within a *personal* framework rather than as a discipline or practice, we can consider each of the recent changes just mentioned—the role of women, consulting, managerial needs, Internet, and biotechnology—from the perspective of how they impact on professional behavior and on other science practitioners. To do this, even in an introductory manner, we need to look at the new ingredients that have been or will be added, how personal relationships will be modified, and how the rules of the game have been or will be changed.

1. WOMEN IN SCIENCE

Women scientists have moved to a remarkable degree toward equality with men in most aspects of their jobs—salary, position, and recognition. Pay scales of women in most scientific institutions are approximately equal to those of men, except at the highest professional levels; academic rank still does not reflect full gender parity, with women often predominating at lower levels; and professional recognition is almost gender-free, unlike some forms of institutional recognition—although even here the inequalities are diminishing. Many younger women professionals state that they have not experienced obvious and deliberate discrimination, even though some may refer to more subtle instances of differential treatment.

What then, in the midst of clear indications of progress toward equality, are the critical persistent problems associated with gender in the science workplace? Three critical issues, each with important personal and interpersonal implications, have been identified: (a) women leave science at a greater rate than men at every career stage; (b) women produce fewer publications than men; and (c) women achieve less, and are recognized and rewarded less throughout their careers, than men. These issues have been explored by sociologists, social psychologists, and feminists (as well as by administrators) for at least four decades, but explanations and solutions have been inconclusive and even controversial. Some insights have been gained, but the problems remain.

   a. *Women leave science at a greater rate than men.* An obvious but too simplistic response to this statement is "Yes, they do, because they have the alternative of creating a family, while men, in their continuing roles as 'breadwinners,' feel less able to make major career changes." Other more thoughtful responses could be that women find themselves in an unequal struggle with male scientists in terms of salary, promotion and recognition, so at some point they simply give up the game—or that some women are unwilling to make the total commitment to science that is demanded for success, so, refusing to accept mediocrity, they quit. One additional response might be that some women feel isolated and deprived of the expected collegiality of science; they often find intense competition instead of cooperation, and hostility instead of friendliness—transforming the scientific community into a cold, unwelcoming male enclave.

   b. *Women produce less, as measured by publications, than men.* Studies of scientists in different institutions and in different disciplines have shown repeatedly that women publish significantly fewer papers than men (averaging only 50 to 60 percent as many) (Cole and Zuckerman, 1984; Zuckerman, 1991). Differences in publication rates begin early in careers of women and increase with age. Furthermore, the relative numbers of women who publish at a high rate are less than those of men (adjusted for total populations of women and men in science). A plausible hypothesis for the observed differences in the rate of publication by women and men as they get older is that "women's access to [research] resources relative to men's diminishes with time, and possibly their commitment to research and publication wanes as they receive fewer rewards and incentives to continue" (Zuckerman, 1991). One favorable sign is that younger women are publishing more than earlier generations did, so the cumulative gender differences will undoubtedly decrease.

c. *Women scientists achieve less, and are recognized and rewarded less throughout their careers, than men.* Visibility and recognition in science is strongly correlated with research performance and publication, so it should not be surprising that women were described two decades ago as "on average, less visible than men; their work is perceived to be of lower quality, and they are rarely mentioned as being major contributors [to their specialties]" (Cole, 1979). This statement would be less supportable today. Insofar as prizes and awards are concerned—and these are not trivial matters to scientists—no clear analysis of gender differentials has been published, but membership in prestigious honorary societies like the National Academies of Science and Engineering, and winners of Nobel Prizes in the sciences, do indicate a remarkable gender disparity. Only about *ten percent* of the National Academy of Sciences is female. And, of the roughly 150 scientists who have won the Nobel Prize over the last twenty years, only *four* have been women. There is, however, one positive indicator. The differential in career attainments between younger women and men is narrowing—a trend which, if continued, will change the future statistics about recognition and rewards for women scientists.

So positive changes have characterized the recent participation of women in science, even though persistent problems exist. From the point of view of interpersonal relationships in the community of science, some important consequences of those improved conditions include the following:

- Today's women scientists enter professional careers expecting a level playing field insofar as gender is concerned, and in most instances they are finding it to be a reality.
- Among younger scientists, a greater attitude of collegiality prevails than was true even two decades ago. Some inequalities still exist, particularly in the mentoring of women graduate students by male faculty members.
- When specific examples of differential treatment because of gender are identified and publicized, institutional remedial responses are usually positive.
- The aggregate of gender-related minor inequalities, such as differential mentoring or lack of adequate professional recognition, may reduce overall satisfaction with a career in science to a level where other job choices are made.
- Women scientists with children—always a stressful existence—are finding greater availability of day care facilities and institutional flexibility, reducing the need for a part-time professional life that retards or derails careers.
- Women are now found more frequently in the managerial and executive levels of science—as program leaders, laboratory directors, or agency officials. At these levels, expertise unrelated to a technical discipline becomes of increasing importance to advancement, when accompanied by proficiency as a scientist.
- Women who are leaders of scientific groups or organizations tend to be very sensitive about gender parity within and outside their perimeter of control. The present status of women in science will be examined more completely in Chapter 10. For now, however, my general observation is that the comfort level between male and female scientists has increased as the disparities have decreased, and the trend is a favorable one.

2. THE CONSULTANT

Large and increasing numbers of the scientifically trained population are involved full-time or part-time in *consulting*. These are people who provide technical advice for a price. They are the synthesizers and interpreters of scientific data, as well as translators of increasingly complex technical information for an expanding population of uncomprehending potential consumers. Most of them are entrepreneurs with advanced degrees in science who have elected careers outside the traditional professional tracks of teaching and/or research. They are supported financially by an evolving information-based society that requires a larger and larger pool of people with the expertise and the temperament to act as intermediaries between those who produce knowledge and those who use it. Such is the primary role of the consultant.

A recent survey of more than one hundred science consultants (Sindermann and Sawyer, 1997) disclosed that a majority of respondents were either satisfied with or were enthusiastic about their jobs. They mentioned in particular the freedom to function as independent professionals—to choose their own work assignments and to decide how their time will be allocated. Some pointed out the reality that it is only in consulting that a scientist is in full control of his or her destiny—that every other job in science is subject to the whims and wishes of some kind of institutional hierarchy.

Scientists may follow several avenues to consulting as a career option. Some elect to join an existing consultancy, or to form a sole proprietorship or partnership venture soon after graduate school; others make a mid-career decision to become consultants after having achieved credibility in research specialties; and still others retire from academic or government positions and immediately begin consulting practices, using contacts developed during their earlier careers to get started. Commitment may be full-time or part-time, and among the part-timers university faculty members are the most numerous.

They exist in every research university in the country—scientists with technical specialties that make them attractive to industry as paid consultants. They meet their classes, attend committee meetings, give seminars, and mentor graduate students, but occasionally they disappear from the campus for a day, or three days, or a week, to travel to some distant industrial park for meetings with corporate officers and corporate scientists. The industry people are looking for new ideas, techniques, insights, concepts, or syntheses that may be useful in product development, and they are perfectly willing to buy relevant academic expertise—usually at bargain prices.

Most of these university-based, part-time consultants are pleased—as well they should be—with an arrangement that provides substantial income and/or stock options to augment a perennially meager academic salary. They are able to live in two professional worlds, each remarkably distinct from the other. Academia is a comfortable place of traditional values and colleagues who are also friends; industrial research labs are product-oriented, high-pressure places usually peopled with very bright, aggressive, journeyman-level scientists. Involvement in a schedule providing a mixture of the two environments can be intellectually stimulating as well as career enhancing.

Arrangements that lead to part-time consulting by university faculty members often evolve slowly. Some factors include (1) rapidly developing competence in a research specialty that has obvious or potential commercial applications (such as immunochemistry, genetic engineering, biotoxins, toxic waste inactivation, etc.); (2) publication of a series of very good research papers in that specialty in prestigious journals; (3) participation as a speaker or organizer in workshops and symposia in the specialty area; (4) publication of a review paper dealing broadly with the specialty area; (5) participation in scientific societies whose membership includes a significant number of industrial research scientists working in the specialty area; (6) correspondence with and visits to industrial research scientists; and finally (7) submission of proposals for research contracts to several companies known to have product development interests in the specialty area. Thus, a blend of competence, visibility, and communication can place an academic scientist with the right research interests in position to become an industry consultant. It should be clear that the results of basic research can be of interest to major production companies, as well as results of applied work, and often fundamental studies can be the principal contribution of academic scientists.

The consultant population is large and diverse, but it is possible to discern a number of guiding principles:

- The conversion from "scientist" to "consulting scientist" constitutes a major career change, motivated principally by (1) a design for freedom to work as an individual, (2) a need for opportunities to utilize creative talents, (3) a search for roles that maximize qualifications and expertise, and (4) interest in financial well-being. Consultants are important in a world identified today by rapidly expanding technology. They serve as analysts and interpreters of scientific data, and act as advisors to clients who need information in understandable form.

- Consulting is a business; recognizing this fact, a science consultant must attain high levels of professionalism and expertise as a *business person* as well as a *scientist*. For the new consultant there are strategies and tactics drawn from both disciplines that promote survival first and then success. Prominent among them are adaptations to administrative and financial matters important to business success, combined with maintaining competence in a technical specialty. Of importance also is the development of a code of ethical conduct that includes both sides of the venture—scientific and commercial.

- The future of science consulting in an information-based society seems assured. The rates of technological advance and acquisition of new information will require the presence of experts to digest, assimilate, synthesize, and interpret information for populations of politicians, executives, managers and bureaucrats (as well as the general public) who will have little understanding of the science involved. Consulting, therefore, becomes a career choice that ranges alongside the traditional ones of research and/or teaching.

From the somewhat restricted viewpoint of interpersonal relationships, some of the important added elements here are:

- In becoming full-time consultants, scientists must distance themselves from research and publication, except for those activities that are necessary to fulfill contracts.
- Full-time consultants have many business demands in their workday that often result in decline in general scientific competence, except in very narrow areas of expertise important to the consultancy.
- Consultants who do participate in professional activities of their disciplines may be seen as non-contributing members, especially if they make too much use of the term "proprietary information."
- Consultants have been accused sometimes of sacrificing scientific objectivity in interpreting data in favor of clients, even though most consultants deny this vehemently.
- Consultants who are also university faculty members often spend significant amounts of time off campus and are unavailable for student contacts.
- Networks of consulting scientists overlap those of academic scientists only partially, and access is often limited. Some academics look on consultants as parasites and deliberately exclude them.

It seems that in becoming full-time consultants, scientists are transported to a different universe closely allied to business. The transition is difficult, although the rewards may make it worthwhile.

3. THE SCIENCE MANAGER

Science today is becoming to a greater and greater extent a cooperative activity, practiced by *teams* of compatible professionals and support staff. The size of the unit may vary, from a single project to a research center consisting of a number of laboratories. At each size and level of complexity one scientist will usually be designated as "principal investigator" or "program leader" or "manager" or "director." That person will be responsible for the direction and focus of the research efforts, for deployment of funds, for recruiting and motivating staff, for effective communication, for productivity and the quality of the work done, and for acting as an authority figure in the discipline area of the research organization that he or she heads—a very large order.

Done well, there are few activities in science that are more satisfying, more rewarding, or more ego-enhancing than leading an effective research team. The group, usually consisting of other professionals, technicians, and aides, plus graduate students and post-doctoral fellows, can be variable in size and complexity depending on the nature of the research. The professional staff is often multidisciplinary, with representation from two or more of the major fields of science. The team may be created *de novo*, with a leader designated by higher authorities, or it may evolve gradually by accretion of supporting staff when a research project or laboratory receives larger and larger funding. Expansion is invariably accompanied by many forms of stress—for the principal investigator and for other professionals in the team. How these stresses are met and dealt with helps to determine the continuity and to shape the productivity of the group, so the kind of leadership that exists is of course a critical consideration.

The leader of an expanding research team must, at some uncertain staff size, begin to recognize himself or herself as a "manager," even though he or she may try desperately to remain actively involved in some phase of the scientific work. Problems

of recruiting, purchasing, public relations, funding, and personnel occupy much of the time that is not spent in committee meetings or in travel status, gradually squeezing out of existence any
time for hands-on research. This is the reality that science managers face and eventually have to accept, recognizing that other
professionals on the team will do the research and publish the
technical papers, and that *"managing for elitism"* in science is a
worthwhile career goal and one with great satisfactions.

The concept of managing for elitism deserves further exploration in a book with "winning" in its title, since this is what
most science managers attempt to do, with variable success. It
can be a somewhat nebulous concept, though, probably best
approached by discussing some of its elements:

- A basic dictum of elite science is *"Do research that matters"*—that addresses large, basic questions rather than
  narrow, mundane ones, that uses multiple approaches
  with current technology, and that is quantitatively im-
  peccable.
- Another dictum is *"Select the best available talent for the
  team."* Choose people with intelligence, energy, and
  training, and then provide the kind of environment that
  will stimulate their productivity and encourage their
  retention.
- Still another dictum is *"Select the most effective supervisors
  and create a functional hierarchy of authority"*—but then
  provide counseling and training to ensure growth in
  each position.
- Beyond these admittedly idealistic dicta, there are other
  keys to managing for elitism. One of the most important
  is *managing for financial stability*. This is a particularly
  acute need in scientific organizations supported by
  grants and contracts, and it is one that requires great skill.
  Sequences of large and small grant proposals, and over-
  lapping but non-duplicative submissions must be
  orchestrated—together with constant searches for new
  funding sources. Good scientists on the team expect that

the institutional funding will not fluctuate widely, so program continuity can be maintained.

- Another key to building elitism is to manage an *"open organization"* in which visiting scientists are welcomed, joint research with other investigators is encouraged, released time is available for specialized or refresher courses, scientific society meeting travel and paper presentations are expected, professional staff members are encouraged to teach a course in their specialty at a local university and to accept adjunct faculty appointments, and an active seminar and conference schedule is scrupulously maintained.

- Still another management practice that encourages elitism concerns the *equitable determination of credit and authorship for results of completed research* to which several scientists have contributed. Few subjects are more important to individual research people than use or abuse of their *"intellectual property."* Clearly stated and mutually agreed upon organizational guidelines should be available for determining authorship, for assigning credit in administrative documents, and for responsibility in manuscript preparation.

- Then there is the poorly defined but very real entity called *"esprit de corps,"* that characterizes most elite research groups. Some of its ingredients are a sense of active participation in a worthwhile venture, respect for the competence and productivity of coworkers, a relaxed work environment, but one with high expectations, and administration by an intelligent, credible scientist/director who treads softly but firmly.

From the viewpoint of interpersonal relationships, some of the most critical elements at managerial levels of science administration are these:

- It is difficult for managers to maintain competence in the increasingly complex disciplines of science and to function as effective leaders of large organizations. This is a reality that leaders must face as the team expands.

- A corollary to the previous statement would be that the science manager must depend on professional credibility as well as on managerial skills—unlike the industrial production manager, who depends principally on competence in analyzing problems and achieving successful solutions. The science director must continue to speak and act as an authority figure in the discipline area of the research organization that he or she heads.

- Managers must be sensitive to the internal power structure of their organization—and must ensure that its boundaries are limited and that it does not conflict with designated responsibilities and authorities.

- Effective routes of communication become ever more critical as a scientific organization becomes larger and more complex. Links must be installed and maintained that are vertical and lateral and that include all members of the team.

- The leader of an expanding scientific organization must be politically sensitive—he or she must be able to interact successfully with politicians and the public, as well as with the heads of other scientific, regulatory, or advocacy organizations.

- The leader of an expanding scientific group must grow in administrative competence—which should be a combination of innate abilities and learned activities. Professionalism must extend to management methods as well as science, and should include formal training in areas such as management practices, psychology, and sociology.

- Science managers, for their own psychological well-being, should try to understand and accept the transience of power. Loss of power, by demotion, displacement, or forced retirement, can represent an overwhelming blow, especially for people who confuse their positions with themselves.

So the managerial scientist is very much a phenomenon of the present era of team research and large research organizations. Professionalism in managerial skills can be expected from such individuals, as a critical augmentation of scientific credibility.

4. THE INTERNET

Information transfer among scientists has already been greatly enhanced by electronic communication—e-mail and the Internet in particular. E-mail not only reduces the time involved in sending and receiving messages and data, but it also facilitates simultaneous correspondence with clusters of colleagues—a method of planning meetings, improving draft reports and manuscripts, or even refining a multi-authored paper when participants are dispersed geographically. The Internet, and its offspring, the World Wide Web, can perform truly amazing feats in fostering communication among scientists. Research findings can be reported, updated periodically, and made available in definitive form. Some of the other almost endless uses of the Internet by scientists include these:

- Awareness of new developments in any specialty area can be maintained by scanning home pages of selected colleagues, research organizations, science publishers, and scientific societies (this may include access to abstracts or summaries of papers presented at professional meetings, workshops and symposia, or, in some instances, entire but unreviewed technical papers).
- Some enthusiasts announce loudly that science is moving toward all-electronic publication—to a day when information is immediately and globally accessible to all, and when print journals will become obsolete. Such a state of affairs does not exist now. Problems in achieving it abound. A large one is the future role of traditional

libraries, which might be transformed into information and archival centers. Another question concerns the survival and continued functions of reviewing and editing —significant ingredients of current science that may fit uneasily if at all into a future characterized by instant publication vehicles such as "e-prints."

- One very powerful arbiter of future development and use of electronic communication in science is and will continue to be *cost*—cost to the individual scientist, to system servers, to software producers, and to journal publishers. The cost factor seems to predominate, for example, in discussions about the future of print journals and payment for online access to the technical literature.

It should not be surprising to find that minor shreds of uncertainty exist as scientists rush to exploit the advantages of electronic communication. After all, the World Wide Web was not initiated until December, 1990, and the Internet became a reality only one year before that (after some 20 years of development as ARPANET by the U.S. Department of Defense, and as CERNET by the European Particle Physics Laboratory [CERN] in Geneva, Switzerland) (White, 1998). E-mail and the Internet have already made significant changes in how scientists communicate and stay current with the action in their specialties, and the technology is moving fast.

Computer-assisted communication is now almost a universal component of most scientists' working day. Information and requests for information move instantaneously on a World Wide Web whose dimensions expand daily. Participation in the benefits of this information input, transfer, and output system is accessible to all, but is subject to some *suggestions and admonitions*, especially those that involve interactions with peers and colleagues:

- Read any Internet offering with a liberal amount of *skepticism*, unless the information can be substantiated or validated. Always remember that the Internet in its present form permits almost anyone to distribute unedited information on any subject—correct or incorrect—widely and inexpensively.

- Be sure to distinguish *recreational* versus *professional* uses of the Internet, and never, never during working hours, become a bored surfer without focus, " . . . wandering the Web from one node to the next, glassy-eyed, weary, waiting to be entertained" (Hayes, 1994).

- Establish and maintain a personal or organizational home page on a topic of continuing interest. The status of research programs can be summarized for colleagues and updated periodically (this can be a time-consuming project, if it is done effectively, but it can be a great networking aid, and a career-enhancing one as well).

- Communicate current research activities and research findings to a very carefully selected community of specialists in a precisely defined subdiscipline of science via the World Wide Web. Use the Web for queries which may occasionally bring unexpected contacts with scientists in other countries—contacts that may be expanded into fruitful exchanges and even to joint projects.

- Be very circumspect about using the Internet in attempts to establish priority for any concept, perspective, or observation. "*Intellectual property*" is a valued component of any scientific career, but property rights should be established in the traditional way—publication in a peer-reviewed journal.

- Participate in discussions about all-electronic publishing, since decisions by scientific societies and commercial publishers may affect how interactions with colleagues on technical matters will occur in the future. Some issues—such as the way that publication will be financed—are still being debated, in this time of possible transition from print journals.

- Follow a "code of good practice" for e-mail use. Avoid trivia—especially if it extends beyond one sentence; and don't "spam" (addressing one e-mail message to many recipients, whether they want to be targets or not).

• Be very judicious about allocation of time between networking on the Internet and research activities. Both can be fascinating; the Internet can be seductive, but research should have priority.

The Internet is clearly a formidable tool for communication among scientists. Still beset by problems, such as lack of any restriction on content, it is, even in its early phases, serving a vital global function.

5. BIOTECHNOLOGY

A recent (January, 1999) issue of the news magazine *Time* carried fifty pages of well-researched articles assembled under the umbrella title "The Biotech Century." Focused particularly on the future of medicine in the 21st century and beyond, the series included summaries of progress in areas such as DNA mapping, gene therapy, and the role of genetics in pharmaceutical development. The central point was made early in those pages that physical scientists have had their way with the 20th century—atomic fission, the evolution of computers, space adventures and exploration—but *there is every indication that the next century will be characterized by the almost limitless development of biotechnology*. If this prophecy proves correct, the image and practice of science will change drastically, and the principal actors will be those with specialties such as informatics and proteomics that are only now beginning to emerge.

The first biotechnology milestone of the new century is a *description of the human genome*—a massive multi-institutional government-funded project now in progress, aimed at determining the biochemical code for all of the +100,000 genes, and then how each one functions, normally and abnormally, in health and disease (at present, about 4000 heritable diseases of humans have been recognized). This still-incomplete program is already being assaulted by a rush of profit-driven private ventures to patent those parts of the genome of greatest medical significance (and economic value)—even before the government-sponsored project is completed. The broad zone surrounding the understanding and manipulation of the human genome—generally identified as *genetic engineering*—allows us to think of conquest of cancer, growing new heart blood vessels, creating replacement organs from cultured stem cells, and retarding cellular aging changes.

The present focus for biotechnology is human medicine, in particular *prenatal genetic testing* for chemical indicators of certain disorders and inherited diseases; *gene therapy*, involving the still-experimental introduction of beneficial genes into the cells of patients; and *pharmaceutical development*, concentrating on target chemical receptors in body cells and the identification of modifying chemicals specific for those receptors.

Now, admittedly, this perspective on the future of industry science, focused as it is on biotechnology, might be discounted as too narrow by professionals in other specialties of science — and yet a little introspection can yield insights about the pervasiveness of this new technology. We talk here about mapping the human genome as a work in progress, but what about the much more difficult task of mapping the neurons of the human brain—leading toward creation of artificial intelligence and the replication of the human mind in a machine—a project that could engage the best minds in every scientific discipline (as was suggested in the same *Time* report).

We should also consider the longer-term role of genetic engineering in food production. Some limited progress has been made with gene insertions to produce plant strains resistant to low temperatures or to droughts, or strains with built-in natural pesticides, or strains that have higher yields of proteins. Attention is also being directed to plants tailored to give large yields in marginal lands. Limits are only those imposed by the human imagination, tempered, we hope, by some measure of conservatism and by continuing ethical scrutiny.

The new century promises to bring science-driven major changes in human existence—changes that are already beginning to emerge under the broad umbrella of "biotechnology." Genetic engineering will be the core activity for advances in medicine, pharmaceuticals, and food production. Those advances will undoubtedly shift the overall emphasis of industrial research.

Many of the investigators in these new high technology ventures will live and work in a different research environment and will have characteristics different from their colleagues in more traditional specialties:

- They will speak in a unique jargon that is almost incomprehensible, even to colleagues in closely related disciplines. They will consider the patent process as an essential feature of their research, with patent applications having precedence over any public releases of research results.
- They will participate more and more in closed workshops and invitation-only conferences, carefully avoiding the large general scientific meetings.
- They will make less and less use of print journals, substituting e-prints and e-mail for information transfer to selected peers in elaborate, closely controlled networks.
- A large percentage of them will be employed by small, high-risk start-up companies, many of which will be swallowed by large, multinational industries.
- The claim of "proprietary information" will become a larger and larger inhibitor of free technical discussion (or publication) in many specialties.
- Many will be employed in medically oriented research on applications of genetic engineering to the human genome—especially in disease control. Applications of genetic engineering to food production will also be a

major focus of research funded by public and industrial sources.

Biotechnology, then, is an early frontrunner for major impact on science of the early 21st century. Sensible game players with proper credentials will position themselves to participate fully.

The five areas of change in the relationships of scientists with each other and with the rest of the world that have just been discussed—women in science, consulting, managing, Internet, and biotechnology—must of course be accompanied by descriptions of the more traditional aspects of professional interactions. Those stable components, listed earlier in this introduction as chapter headings, will occupy us for the rest of the book; they constitute the foundation for the stage on which ritual scientific games are performed.

I invite you to join me onstage. Stand with me on some of the sets; tiptoe through controversial scenes; reject some advice and conclusions as sheer hogwash; accept and enjoy an interlude here and there; agree or disagree with any of my premises; and, finally, give me some of your thoughts on this fascinating topic of *Winning the Games Scientists Play.*

A PRIMER FOR SCIENTIFIC STRATEGISTS

So much of what we call "science" consists of people interacting with people. Nowhere does this become clearer than during an attempt to identify and describe some of the things that scientists do (other than to produce and analyze data). Part One of this book considers some of the more important of these nonlaboratory activities, including writing and presenting papers, attending meetings, chairing sessions, organizing conferences, and participating in committee meetings. All of them require interpersonal skills—the kind that develop from awareness, study, and practice.

The thesis of Part One is that the joys of science can be much enhanced if its practitioners are knowledgeable about people as well as about technical matters. The thesis is a simple one. Why is it so frequently ignored? Part of the answer may be that professionalism per se is not among the course offerings of most graduate schools in science, so awareness is missing. Another part may be that many interpersonal strategies related to the practice of science are subjective and variable, therefore easily dismissed as "nice to know but non-essential." Still another part may be that some scientists are so totally involved with technical activities that they are oblivious to the "people halo" that must surround the physical and mental processes of "doing good science."

Part One attempts to point out the need for effective interpersonal relationships in many of the principal activities of scientists, and even attempts to provide advice about correct procedures (always a dangerous undertaking). It is clearly a "primer," but what more logical place is there to start?

CHAPTER 1

THE SCIENTIST AS A WRITER
Publishing Scientific Papers

Maneuvers associated with senior authorship; how and when to give credit to others; rules of the road in a "publish or perish" environment; reviewing the writing of others; an editorial view of scientific writing.

Scientific game playing reaches its absolute peak in the arena of publication—an entirely logical evolution and a reasonable place to begin this book, since one of the essential components of scientific research is the formal presentation of data and conclusions for critical scrutiny, evaluation, and testing by peers. Central issues include authorship, priorities of research, satisfaction of administrative "publish or perish" dicta, interaction with journal editors and reviewers, and establishment of credibility in the scientific community through publication. These and other topics relevant to publication are included in this chapter, which is innocuously and possibly incorrectly subtitled "Publishing Scientific Papers." The *methodology* involved in writing good scientific papers is not and cannot be a primary objective of this chapter, since books have been written on the subject. The assumption must be made that the "how to" aspects have been or can be acquired from such publications, from instruction at the knee of a major professor, or from the red-pencil comments of tough-minded reviewers and journal editors.

In the following sections I plan to approach scientific writing from three viewpoints—those of the writer, the reviewer, and the editor. Then I will consider some exciting new concepts in paper evaluation.

AUTHORSHIP OF SCIENTIFIC PAPERS

Authorship of original, sometimes creative, scientific papers, and of reviews and books, is one of the most important satisfactions of a career in science, paralleling and sometimes surpassing satisfactions of peer approval and salary (I said "sometimes"). The pleasures of sole individual authorship are sweet indeed, but except in the case of the autonomous independent investigator such pleasures are too simple to be real. The reality, in an age of complex, often interdisciplinary research, is that of joint authorship, with its associated hornets' nest of problems.

First authorship—the name which appears first in a list of two or more authors—is of course the plum, especially in multiple-author papers where *all* the other authors become part of the "et al." and essentially disappear from view forever. First authorship of a paper must necessarily be "deserved," but the definition of that word depends almost totally on perceptions. To the senior faculty member or laboratory director the definition is simple and clear-cut. To the junior faculty member, or member of a research team, who may have contributed significant ideas and insights as well as most of the laboratory work, the definition is definitely not clear-cut. Even the technician, who may feel that he or she has contributed substantially in terms of planning as well as execution of the research, may not find the issue to be crisp and simple.

In the face of all these varied perceptions and personal perspectives, a clear and universally accepted set of rules would be desirable. No such rules exist or are apt to exist, but here are ten possibilities:

1. *A critical, never-to-be-overlooked aspect of joint authorship must be early and unanimous decision, even before any collaborative research is started, about first authorship and even about the sequence of authors on the published paper.* More dissension, unhappiness, bitterness, hostilities, and long-standing enmities have been created by failure to observe this basic rule than by almost any other single issue in science.

Properly, the person who has contributed most, in terms of original perception and definition of the problem, conception of the research required, detailed description of research protocols, analysis of data, and formulation of conclusions, should emerge as first author. What could be simpler? Of course, in a joint research effort different individuals may have dominated particular phases of the research—and herein lies the problem. A simple solution, and one frequently proposed, is *that the person who originally formulated the research concept, and whose original insight was responsible for the study, should emerge as the first author, regardless of the subsequent contributions of others.* This proposal would obviously run a high risk of unacceptability to those who carried out most or all of the actual research—either as peers or junior scientists. Despite this risk, it seems like a logical basis for discussion. The early maneuvers for first authorship, and possibly the final decision, may focus on this concept.

Probably the significant point here is that a good game player will not become involved in a multidisciplinary study with multiple authorship papers without knowing in advance what the game rules are.

The debate about correct determination of the sequence of authorship has dragged on for decades. Solutions which have been proposed include approaches other than the one mentioned above—that the per-

son responsible for the original concept and initial planning should emerge as the first author, regardless of the subsequent contributions of others to the work on which the paper is based. At a late-night small-group conference in a New England hotel room, the following somewhat fuzzy alternative proposals were suggested:

- If the concept and planning of the paper are developed in group discussions, first authorship can be decided by a toss of coins, with all the losers being listed alphabetically, or listed on the basis of subsequent coin tosses.
- If the research is supported by grant or contract funds, then the designated principal investigator should be the first author. If it is supported by two more or less equal funding sources assigned to different investigators, then we're back to the coin toss.
- The principles of chivalry might be invoked to decide first authorship. Female scientists would be easy winners here, as would scientists of advanced age or physical frailty. (This proposal was received with a noticeable lack of enthusiasm by the group.)
- The approach of choice—elucidated just before the group was dispersed by the hotel management—turned out to be that if the research supporting a proposed paper has been started or completed without a decision about first authorship, then the sequence of authors should be determined by a complex formula, whose major elements are:

  Conceptual input (C)
  Planning input (P)
  Data acquisition (Dac)
  Data analysis (Dan)
  Hours of time invested (T)
  Preparation of first draft (Pd)
  Final editing (Ed)
  Each element is weighted, and the formula becomes:

$$\frac{4C + 2P + 2Dac + 2Dan + Pd + Ed}{T^1}$$

Assignment of values for each element is done by a select committee of peers. Differences among total scores must be tested for statistical significance, and where significance cannot be demonstrated, additional factors should be included in the formula. These are:

> *Years since receipt of Ph.D. (Sr)*
> *Comparative sizes of relevant research grants (SS)*
> *Academic rank (if university) or GS rating (if government) (R)*

The formula for determination of first author provides some protection for the concept originator, but also gives strong contributors to other elements involved in the research a fighting chance at the prize. Results of initial tests of the formula with actual manuscripts can best be described as "inconclusive."

The role of technicians in research and publication must be clearly defined in every group. Normally technicians are instructed in what needs to be done; they then do it—and if they do it conscientiously and diligently they should be mentioned in an acknowledgment section of the published paper or in a footnote. Only if they contribute significantly to the thinking underlying the research should technicians expect any more than this—and even then the decision about possibly including the technicians' name as junior author is completely and absolutely the prerogative of the scientist. There is, however, a broader principle here which may be overriding. Good game players are careful to give adequate credit and a piece of the spotlight to associates and assistants—without disappearing into the scenery themselves.

2. Regarding the *citation of the work of others* that is directly

relevant to the research to be reported, there is an obligation to recognize substantive closely related studies. Beyond this there is wide and legitimate latitude for the scientific game player. Complete cynics would subscribe to the following principles: (a) minimize where possible (but do not ignore) the literature contributions of those who have incurred your dislike, for whatever reasons; (b) maximize, to the extreme limits of modesty, your own literature contributions and those of your close associates in other laboratories; (c) be prepared to point out glaring deficiencies of other studies competitive with your own, in whatever forums that are available; (d) slash or at least slightly wound the competition in introductions or literature review sections of papers; and (e) make it clear that the "others" are interlopers and usurpers, deficient in data, naïve in conclusions, and sterile in concepts. Game players are cognizant of these cynical principles, but avoid their harsh application in scientific writing—preferring moderation and detached professionalism, devoid of personal considerations.

3. With reference to the work of others, there is a special class of scientific writing that has virtually disappeared from the pages of most journals—*comment on or disagreement with published work of others*. In the earlier, simpler days of science it was not uncommon to have running commentary between individuals or research groups. Within reason, this seems like a fruitful way to exchange ideas, to present new data, and to voice disagreement in writing. Today this outlet has been reduced to rare letters to the journal editor pointing out serious discrepancies in something printed in the most recent issue of the journal. The American journal *Science* and several European journals are among the few that persist in providing such a forum. Because of the virtual disappearance of this method of communication, only one other avenue exists for correction of seeming errors in methods or interpretation of others—this is to include some mention of the deficiencies in the introduction or discussion of a paper published by the dissenter, or in a review paper. Usually these alternatives are unsatisfactory, since the disagreement

must be summarized very briefly to avoid disrupting the trend of the new paper. The result is that in most instances the apparent errors in published papers go uncorrected or unchallenged, unless the author chooses to point them out in a subsequent publication—which is also a rare occurrence.

4. *Acknowledgments* are often treated too lightly by authors of scientific papers. It costs little, and the rewards are substantial, to acknowledge the often-dedicated work of technicians, aides, secretaries, illustrators, and even students who have contributed to the final publishable product. As a rule of thumb, all those who have contributed to the research beyond what might be routinely expected from reading their job descriptions should be mentioned in the acknowledgments section. Acknowledgments should be direct and sincere.

Avoid extremes of effusive praise in acknowledging the help of others, and be sure to give disclaimers to protect reviewers, if they are identified (and they should be asked if they wish to be). Remember that they may not want to be associated with the paper; and even if they do not object to being acknowledged, they may not subscribe to your conclusions, in whole or in part. Reviewers may even sue you if you list them against their wishes.

An extreme example of the pitfalls surrounding acknowledgments was related to me recently. Involved was an aggressive graduate student at a large eastern university. The student was impatient with progress on his dissertation under his thesis advisor, and was reluctant to invest any substantial amount of time in further research (in part because of numerous outside commercial interests). The student, a good salesman, was able to convince another professor to act as his thesis advisor, and, quickly submitting his dissertation, was granted a Ph.D. Unfortunately, in the acknowledgments section he gave rather effusive praise to his original advisor, who, when he became aware of the "recognition," immediately filed a civil suit against the new Ph.D. for defamation of character, claiming that unde-

sired association with the thesis impaired his (the professor's) scien-tific credibility. The case was heard, and though no damages were assessed, the professor had his public forum, and the student learned a very painful and indelibly imprinted lesson in human relations.

---

5. *Publication in a research area undergoing explosive expansion can be a very delicate operation.* At any given time there will be a number of narrow research areas that are being explored vigor-ously by several research groups or individual scientists. Exam-ples might be the rapid development of theories and informa-tion about the biochemistry, mode of action, and transfer of genetic material, or earlier work with adaptive enzymes in bac-teria. Publication and establishment of priorities of information in such research areas often become quite competitive. Devices used to establish priorities (properly called "territorial limits") include distribution of preprints (sometimes privately pre-pared) of material in advance of publication, disclosure and publication of information in newspapers, or use of house organs or captive journals which give preferential treatment to one group or another.

Unseemly squabbles can develop when similar ideas or approaches occur to and are pursued by several groups or indi-viduals simultaneously. Occasionally, and unfortunately, these minor tempests can reach the public in various ways—and this tends to belittle science and scientists.

A classic example of the realities of the personal infighting and downright nastiness that can result when scientific groups compete in a rapidly developing area was published in a newspaper feature story. Reference was made to gene splicing by a university group, but much of the report dealt instead with the maneuverings of the principal com-peting scientists to score before the competition. All of the classical devices common to such squabbles—snobbery over presence or absence of degrees, competition for laboratory space, early release of findings in forums other than scientific journals, secrecy about research projects underway, removal of experimental material by an investigator who defected to the opposition camp, applications for patents by those only peripherally involved in the actual experimental work, and threats of lawsuits—were laid out in detail. The average reader must wonder how any progress is made in such an atmosphere, and must conclude that scientific minds can be remarkably small and petty on occasion. Scientists, after all, are people, too, and are not entirely free from frailties of the human character. The article should be required reading for beginning gamespeople (Stockton, 1980).

6. *Additional rules of the road are necessary in a "publish or perish" climate*—the existence of which administrators deny but the reality of which is often clearly demonstrated. Administrators are impressed by numbers and weight, so rewrites of the same material under slightly different titles are in order, as are impressive covers on reprints and the citation of abstracts, notes, and book reviews in an annual list of publications.

There is, of course, some basis for debate about the ethical aspects of rewrites. If they are designed for widely different outlets—such as the popular or semipopular press, then there should be little problem since such outlets require extensive rewording and usually changes in titles and illustrations. If substantial new data or new insights are added, there is also no problem. If precisely the same material is used or proposed for publication, except for minor changes in introduction and summary, then there is a problem. Journal editors can become unhappy, if not unpleasant. Colleagues too can get annoyed and disenchanted if material that sounds faintly or strongly familiar from earlier publications by an author reappears too many times, either in writing or presented orally, using the same old 35-mm slides at meeting after meeting.

7. *Reviewers' comments* on manuscripts should never be taken lightly. Though confusing, ambiguous, and sometimes the products of small minds, such comments occasionally contain a gleam of insight or a nugget of wisdom that should not be rejected out of hand. While any author is free to ignore such

comments, this is done at some peril, and is not done too often or indiscriminately by the proper game player, since in the end you are in the hands of the journal editor if you wish to publish in that particular journal, and since most editors depend heavily on reviewers' comments in making a "go" or "no go" decision.

Actually, reviewers comments can also be used as a learning experience beyond the confines of any particular paper. The game player evaluates individual approaches to reviewing (gentle, uninvolved, nit-picking, violent, helpful, etc.) against the time when he or she will be asked to review manuscripts. Scrutiny of comments in this detached way can help to lower the blood pressure and can aid in the formulation of a philosophy of reviewing. Possibly in this way some of the extreme forms of reviewing could be softened. Harshness and sarcasm could be avoided, and the "piranha technique" of slashing a paper to ribbons without making the author aware of attack would be more and more unacceptable, unless the manuscript obviously attempts to ride on an inadequate data base or is really poorly written; in this case it deserves the worst.

As a general operating principle, authors should consider *all* reviewers' comments very carefully, revise the manuscript in accord with those that result in improvement, but not feel constrained to accept every last comment and to follow slavishly every suggestion. There is room here for discretion. *However,* if major comments by reviewers are not accepted and incorporated into the manuscript, the author should explain clearly and completely in a letter to the editor why they were not accepted. Editors are often reasonable people, but if a choice exists the editor usually will favor the reviewer and not the author.

There are other positive ways of coping with reviewers' comments. The author should accept the fact that there is no substitute for quality of scientific data and conclusions, combined with writing skill. These are the magic ingredients that every reviewer would like to see and that every author would like to produce. However, some if not most reviewers and editors have "pet peeves" or particular likes and dislikes that are reflected in reviews. Some reviewers will always find fault with the structure of a paper, some will always insist on alternative statistical treatments, some will complain that sentence structure is too complex, some will want additional experimental data, on and on. A single bad review should not be considered traumatic or should not elicit thoughts of suicide—you may have merely irritated an editor or influential reviewer with minor and easily corrected deficiencies in style. However, several serious criticisms by different reviewers and editors cannot be dismissed on this basis, and should provoke some private soul-searching.

8. *The apocalyptic day may arrive when, despite the proper deployment and use of game-playing techniques, a manuscript is rejected.* Responses move through a predictable sequence of disbelief, anger, depression, and urge for rebuttal and revenge. The proper game player, however, expects and even relishes this seeming catastrophe because, if he or she is a rule-follower, he or she will know what to do:

a. Prepare an extensive, witty, thoroughly nasty letter to the slow-witted and imperceptive journal editor—but never, *never* send that letter;

b. Examine the rejected manuscript coldly, in the light of the editor's and (more importantly) the reviewers' comments, and ask the gut question—"Are they right?" Then

c. Quickly have the manuscript retyped in the format of another journal after having made changes suggested by the reviewers, and submit it there in supreme confidence that it will be accepted. (It may well be; but if not, after two such attempts, file the manuscript.)

It is important to know that each journal has its own policies on the kind of manuscripts it will accept, and it may well be that rejection is a result of those policies and is not because the manuscript is poor. Sometimes, though, the "policy ploy" is used to reject a poor manuscript without the necessity of the editor's saying it is just plain poor. A second rejection by another journal would provide the answer in such cases.

It should be noted too that editorial boards of some journals can be closed circles, in which the editor selects those of like mind to serve as reviewers and in which members nominate compatible new members as replacements when terms expire. Individual authors should not attempt to feud with such circles; there are too many other journals to be considered.

9. *A day may arrive for many scientists when it seems appropriate and fitting to produce an extended review or even a book.* The decision to produce such a document should not be made without careful introspection, since it has impacts on productivity (in terms of active research) and on family life. There is, however, a time and place when many scientists must, for their continuing peace of mind, make such a decision, and to hell with the consequences. These chosen few must be equipped with a whole new set of game techniques that are foreign to any previously learned.

An early and sobering experience is dealing with book publishers, who are abundant, but who have a universal disdain for the business sense of most scientists. Book publishers are business people; scientists are often patsies. The most important step a scientist can take, after deciding to write a book, is to consult with several colleagues who have published *more than one book,* and *to take their advice* about publishers, advances, royalties, reprint rights, copyright sales, paperback rights—all the nitty-gritties that are important but usually overlooked by the first-time producer of an epic scientific work. Otherwise, the scientist may sign an unfavorable contract, with too low a price on what will prove to be the most time-consuming, exhausting, frustrating, but personally rewarding experience in any scientific career. As in any business field there are jackals in scientific publishing eager to capitalize on the idealism and naïveté of scientists with something substantive and extensive to say in print. These fringe publishers should be avoided, but even then the road is not clear. Some scientific publishers with the largest sales volume and the largest advertising budgets are still not above taking financial advantage of an unsuspecting and uninformed author.

A book is of course the literal and figurative heavyweight in the kingdom of "publish or perish." But writing a sole-authorship book is a difficult, time-consuming, and lonely business. Fortunately there are less arduous routes to book production. One is to invite several colleagues to write a multiple-authorship book for which you will be senior author, of course. Another route that is used by many expert game players, one that has become increasingly popular in the past decade, is to *edit* a book of contributed papers by a number of authors—usually but not always as the outcome of a conference or symposium. There is still a reasonable amount of work involved in editing a volume, and you may have to produce an introductory chapter yourself, but the required effort in no way compares with writing a book, and *your name is on the cover in either case.* (I will have more to say on how to become an editor of such a volume in the chapter on organizing scientific meetings.)

10. *Professional writers of fiction and nonfiction often form what are called "support groups"* composed of other writers willing to comment critically but constructively and supportively on one another's work. Such groups often meet regularly to review and discuss pieces of writing by the members and to make suggestions for improvement. We have seen little of this attitude and approach in science, with the possible exception of temporary circles of graduate students in large universities. Otherwise, the usual sequence is the presentation of research results orally in a departmental seminar, where some constructive advice may be given, but where harsh criticism may also be expected. The scientist then retreats to his desk to patch up his presentation in a form suitable for publication. Occasionally, a draft may be floated past a few colleagues, but usually the fragile offering is placed directly in the calloused hands of journal editors and

reviewers, where it may receive abuse more often than praise. Those scientists who can survive in such a rough arena learn from the process, but often after substantial ego deflation and heartburn. How much pleasanter the act of writing up research results would be if scientists adapted the "support group" concept to their own needs.

THE ROLE OF REVIEWERS

Although this chapter is concerned principally with *writing* scientific papers, one of the best available learning devices is to peer into the other end of the telescope—to look at manuscripts from the viewpoint of the journal editor and his reviewers.

Editors of scientific journals are usually reasonably perceptive individuals; they know that their success depends to a surprisingly large degree on assembling a stable of superb reviewers (also called manuscript referees). Since publication is such an integral part of the research process, the assurance of excellence in formal printed presentation of results and conclusions must never be slighted. Judgment of peers is the ultimate test of the scientist; this judgment is best conveyed as comments by reviewers of manuscripts, and as editorial decisions based on those comments.

Because of the pivotal importance of reviews to ultimate success in science, the game player will instinctively acquire competence, and eventually excellence, in reviewership arts. Additionally, because reviews can be sources of power in science, it is clearly an area for attention by good players.

To my knowledge there are no detailed guides for reviewers. Of course most journal editors have developed a set of *instructions* for reviewers, but these are often just lists of criteria—validity of experimental treatment, knowledge of published literature, suitability of tabular material, rendition of illustrations, logical theme development, rational structure of the paper, ad infinitum. The game player must go beyond such lists to delicate, often intangible aspects of the art—to those things that distinguish the portrait artist from the amateur photographer.

Those things which separate the art of reviewership from the mechanical task of examining manuscripts are sometimes ephemeral but are, like all good art, recognizable by true professionals. Some of the trail markers are these:

- Good reviews must be based on genuine competence in the subject matter area of the manuscript; superficial knowledge will quickly become obvious to the editor and the author.
- Good reviews require substantial investment of time; comments based on cursory examination of a manuscript can be misleading, erroneous, or downright stupid.
- Good reviews require exceptional analytical abilities, particularly those that enable perception of major flaws in design, execution, or conclusions.
- Good reviews require utmost diplomacy and tact in handling hopeful offerings of often-sensitive authors; this is particularly important in examining manuscripts from inexperienced writers.
- Good reviews occasionally require gentle but firm and clear statements of deficiencies in manuscripts, and in rare instances, outright rejection of methods, results, or conclusions.
- Good reviews should constitute important learning experiences for the *reviewer* as well as the author; flaws in manuscripts produced by others often mirror those in our own writing.
- Good reviews can be *creative;* often insights provided by reviewers can provide new perspectives on data or conclusions developed by the author of a manuscript.
- Good reviews constitute acts of altruism by the reviewer, but are simultaneously self-serving; they are individual offerings to the development of understanding, but they

help assure quality of the products of science, and they help keep the reviewer informed about new research.

- Good reviews are never harsh or personal or sarcastic or belittling—despite occasional overpowering incentives to use such techniques on writing that is mediocre or poor. Reviewers must always remember that they are handling tender, newborn creations, often produced by a fiercely protective parent or parents.

This utopian landscape of good reviews would be marvelous to behold—if it existed. Unfortunately the real world of manuscript reviewing is less idealistic and is equipped with many sharp edges and corners. The review process is too liberally sprinkled with inadequate cursory examinations, nonconstructive comments, inordinate delays in responses to requests for reviews, and examples of petty turf protection—often as a consequence of inattention to or rejection of game rules. Journal editors battle constantly for more thorough peer evaluation of the products of research, but too often such evaluations are considered low-priority burdens by reviewers.

Book reviews are forms of writing that provide scientists with maximum discretionary elbow room, to be used for good or evil. Published in major journals, such reviews can be obviously laudatory if the book is good, or less than that if it is not—and this is proper etiquette. Unfortunately, there is a small subset of reviewers who feel obligated to shred and then do ritual fire dances on every book that they are asked to comment on *after* its publication. They are probably close relatives of those scientists who make equally destructive comments about technical manuscripts that unknowing editors send them for review *prior to* publication. Taken together, these ego-destroyers are responsible for much gratuitous trauma inflicted on writers—injuries that may heal, but are never forgotten by the victim.

I label the people who persist in causing this kind of harm as "unprofessional professionals." They may spend their entire careers not knowing or not caring about the consequences of their poisonous reviews—and it is all so unnecessary, if only a few game rules are followed. For instance, never, never use subjective judgmental terms like "useless," "misleading", or "not adequate for professionals"; and before sending off a review, ask whether it will damage the self-esteem of a colleague. Apparently these "unprofessionals" have never been told (or have failed to learn) that statements about the work of others can make their points without using brute force; that even bad writing and/or bad science can be reviewed constructively; and that errors can be indicated objectively, without permitting anything that is poor or even mediocre to make its way into the published literature.

Beginning game players should inspect this "literature within the literature" on reviewing, with a view toward greater understanding of how a professional approaches the assignment—and how *not* to do it, as the following very real example illustrates:

A well-liked and knowledgeble professor of biology at a large northwestern university recently published a book on the population dynamics of wild mammals, based on a graduate course she had been teaching for almost two decades. The book, unfortunately, was assigned by a journal editor for review by a senior faculty member at a southeastern university—a person not known for even vestigial flexibility or humor in the affairs of science, especially in his specialty, which was also mammalian population dynamics. The review was vicious, replete with epithets such as "naive," "incomplete," and "superficial." Serious questions that had to be asked about the review by any reader of the journal were "Why cause such unnecessary pain to another professional?" or "Why would a presumably intelligent scientist prepare such an ill-tempered and highly subjective review?" or "Why didn't the journal editor kill the review on the grounds of unprofessional writing?" To me, the most disturbing aspect of the situation was that the reviewer, after spending decades in academic science, had learned so little about managing interpersonal relationships with col-

leagues, and had apparently never heard of, or had rejected, the concepts of flexibility, compromise, and ego-preservation, which are critical elements of the games (including reviewing) that surround the conduct of good science. But of course the review was published, with its virulence intact, and science is a little poorer because of it.

[The "unprofessional professional" exposes himself or herself by other mean-spirited actions as well—such as the unnecessary abuse of graduate students or junior colleagues after oral presentations (unless, of course, such treatment is invited by pomposity, untruth, or poor data) or by participation in the transmission of rumor or innuendo that can destroy the credibility of a colleague.]

From firsthand observation it is obvious that many scientists do not know how to review a manuscript, nor do they know how to receive comments from reviewers. Blood pressure elevations, hours wasted in drafting witty rebuttals to reviewers' comments, strained professional relationships—all are parts of the price paid for ignorance on both sides of the net in this game. Some authors are so traumatized by harsh reviewers' or editors' comments that they never completely recover. Publication for them becomes a lifelong feud with editors, or an unpleasant chore to be delayed as long as possible, regardless of career consequences.

What seems like an unnecessarily harsh review (whether justified or not) can call forth latent devils in many authors. Key epithets cast at reviewers include "unprofessional," "naïve," "poorly informed," "archaic," "unqualified," "ignorant," "abusive," "tasteless." Authors can simply react in blind rage instead of treating the mean review as a learning experience; but caustic reviewers may sometimes have useful comments, even if their game playing is poor.

Often authors are so blinded by the fiendish need to retaliate against negative reviews that they become totally defensive about their entire paper; they may lose all objectivity and ignore or reject even the valid criticisms. This often results in rejection of the paper by the editor, who will (and should) be somewhat protective of his stout band of reviewers—even those who may be somewhat abrasive. Reviewers, after all, constitute the critical outer defense perimeter of an editor's empire, whereas individual authors are expendable—especially those who lose their cool during the assault.

The game player, whether writing or reviewing manuscripts, maintains distance and perspective on scientific paper production. Good scientific writing is creative, but is to a significant extent an acquired skill, requiring genuine commitment and careful study of the best products of other scientists.

SCIENTIFIC WRITING VIEWED FROM THE EDITOR'S DESK

Another perspective on writing scientific papers can be provided by journal editors, those important but difficult power figures who help determine success or failure in science. Publication in major journals is an essential part of growing up in science, so sooner or later (and preferably sooner) every scientist must place tender offerings—the written results of and conclusions from his or her research—on the untidy desk of a journal editor. Editors have many human failings, but they also have final decision-making power over what gets published or rejected. The game player recognizes the realities here, and acts appropriately (interpreted as conduct to ensure acceptance of manuscripts):

- The science contained in the manuscript must be good, even excellent. Principal foci are adequate experimental design, correct statistical treatment, knowledgeable review of relevant literature, conclusions based on evidence presented, reasonable discussion of findings, well-designed tabular material, and professional illustrations.
- The manuscript should be written in excellent English

and framed logically—*it should be able to withstand scrutiny as a good piece of writing, exclusive of the science.* Included would be logical development of data, conclusions, and ideas; correct and simple sentence structure; effective, lucid word choices; use of topic sentences; avoidance of jargon; elegant simplicity of expression— all the proper things contained in countless books on English grammar, usage, and style, but often ignored by scientists in their haste to present results of research. Good scientific writing should be correct, but more than this, it should keep the reader awake and interested, without being overly dramatic.

- The manuscript should be prepared in the standard format requested by the journal (and journals can vary widely in their formats). Sometimes rejections can be based on failure to comply with simple instructions printed on the back covers of journals.

- The manuscript should be submitted to *appropriate* journals. Each subdiscipline has its representation in at least a few journals, and a manuscript sent to an inappropriate journal will only be delayed by several months before being returned with that notation.

- Editors may or may not appreciate gentle suggestions of names of suitable reviewers—but unless you know the policy, take a chance and offer the names of knowledgeable people anyway. Don't try to insert close friends, buddies from graduate school, or former professors, however, since the editor may just be a little brighter than you think.

- *Proofread the manuscript.* It is absolutely astounding how many authors ignore this rudimentary precept. A remarkable number of definitive drafts of manuscripts reach the editor or reviewers without having been read by the author. They may have many typos, or entire lines of type left out, or incorrect punctuation, or misspelled words—all open invitations to quick return of the unread manuscript, or at least an initial negative reaction by people who count.

- If you have rewritten a thesis for publication, do a thorough job of disguising it. The rewrite should be no more than one-third the length of the thesis, with highly selected tables and graphs, and with severely curtailed literature review and discussion. Journal editors instinctively dislike and are hostile toward any manuscript that is even suspiciously similar to a thesis, with its pseudo-authoritarianism, its multiplicity of tables and illustrations, and its pomposity. A device that sometimes works is subdivision of the thesis material, with a fresh approach to the data and conclusions for each segment.
- Avoid series publications. Many journals won't accept them anyway, and the temptation is great to refer to results contained in previous papers in the series, or to cite many or all the previous papers in the series as a means of establishing turf. These devices are so transparent that they are beneath the dignity of game players.
- When comments are received from an editor, don't get defensive or hostile. You are dealing with an extremely knowledgeable, perceptive, but often jaded individual who expects and insists upon excellence (or at least an approximation of it). He or she *wants* your paper to be as good as it can be, and will often spend hours trying to help you make it so. Don't expect a lot of praise, even for a good paper; that has to come after serious scrutiny of the published work by your peers.

Finally, a journal editor of my acquaintance has offered the following potpourri of suggestions for the beginning writer:

- Don't be afraid to write interestingly, but avoid the overly dramatic.
- Use simple sentences. Simple sentences need not be short. Too many short sentences can be boring. Too many

long, complicated sentences can be confusing as well as boring.
- Don't be afraid to write *correctly.*
- Avoid jargon.
- Topic sentences are great ways to begin paragraphs.
- Logical progression of thoughts is an art form worth cultivating.

Just as exam graders at Harvard (Carswell, 1950) have come to know and love the major kinds of ploys used by unprepared students—the Overpowering Assumption (OA), the Artful Equivocation (AE), and the Vague Generality (VG)—so too do reviewers and editors have their own favorites among the gambits used by authors in attempts at painless paper production. Some of the characters in the script are:

- the "jargon maven," who delights in the use of the complex and obscure dialect of his specialty, fully aware that only nine other people in the entire country will understand;
- the "data stretcher," who kneads and plumps thin results into something barely acceptable;
- the "hasty generalizer," who can extrapolate a single pedestrian finding into a major conceptual advance;
- the "conjurer of obscure formulae," who never uses standard mathematical analyses, but always builds complex and obscure "models" on paper;
- the "discussion milker," whose offerings always include too many flights of fancy disguised as discussion of results;
- the "recycling king," noted for extended mileage and multiple papers from every data set;
- the "great emancipator," who can dismiss or disown with great skill any data points outside a predetermined range;
- the "curve fitter," who can, with a given number of data points, find a linear or curvilinear relationship with pos-

itive or negative slope, depending on the whim of the moment;

- the "joyless drudge," who plods through data as though it were partially dewatered sewer sludge; and
- the "nimble mountaineer," who leaps mindlessly from peak to peak on his graphs, rarely exploring the valleys for true understanding.

These, to editors and reviewers, are like old friends, peering up from manuscript pages with sheepish pixie-like grins, a little abashed at being discovered, and daring anyone to be unkind to them.

DAMMING THE PAPER FLOOD

Publication of scientific work would seem to be routine, following discernible principles of good English composition and good science. In reality there are numerous areas of continuing tension and disagreement which may be encountered by many practitioners. Some of these have to do with the amount of data to be incorporated in a paper, the allowable extent of the reworking of previously published data, and the ways in which publications can be used in evaluations.

The journal *Science* (13 March 1981) carried an article by W. J. Broad exploring the interesting concept of LPU (Least Publishable Unit), a euphemism for fragmentation of data and consequent piecemeal paper production—publishing more and shorter papers with more authors. Statistics cited in the article suggest a significant trend toward an increase in the average number of co-authors and a decrease in the average number of pages per paper—a definite trend toward the ultimate LPU.

Regardless of causes (some of which, like increasing page charges and more multidisciplinary studies, are valid), the trend toward the LPU needs to be beaten back. To assist in the process I would like to point out that the LPU has important but underemphasized counterparts—the Optimum Publishable Unit (OPU) and the Maximum Publishable Unit (MPU).

Unlike the LPU and its scrawny data base, the OPU includes an amount of data that can be comfortably encompassed in a reasonable paper of five to eight printed pages of text, with no artificial subdivisions, with conclusions that are few and crisp, and with discussion that is sharply focused and relevant.

Clues that a paper is approaching or has exceeded that discernible but undefinable boundary of the Maximum Publishable Unit (MPU) are several, including an overlong methods section with too many subdivisions, difficulty in maintaining any thread of continuity in the discussion section, and a list of conclusions numbering six or more.

Both the OPU (Optimum Publishable Unit) and the MPU (Maximum Publishable Unit) are quantifiable and can be determined by journal editors and reviewers using a standard equation. Each editor can have printed on the back cover of the journal (or wherever instructions to authors are printed) his or her concept and definition of LPU, OPU, and MPU, possibly in terms of minimum or optimum numbers of data points; minimum, optimum, and maximum allowable numbers of lines per author; optimum and maximum number of figures or tables per author—all with sufficient wiggle room for the editor to accommodate a wisp of new insight or a hint of fresh synthesis, even if other criteria are abused. Values can be calculated by authors before submitting a paper for publication, or by editors after the paper is submitted. Manuscripts whose values fall outside stated limits can be automatically rejected.

Hand in hand with MPU and OPU must go another concept, the MRQ (Maximum Recycling Quotient), a codification of the number of times and different formats in which the same material may be published. Along with shorter papers and more co-authors there is a perceptible ground swell favoring publication of recycled information. Some authors add a soupçon of new information, change the title, change the introduction and conclusions slightly, add some new figures but no new information, and republish in another journal—the possibilities seem endless, until colleagues or reviewers begin to exhibit unhappiness; then the game is finished. Quantification of the MRQ involves development of an equation in which the new information, new illustrations, and additional analyses are compared with the material in the original publication to derive a single number (the MRQ). The journal editor has printed, on the same back page with the MPU and OPU, his or her allowable MRQ. If the MRQ is exceeded the paper is summarily rejected. Consistent violators of MRQ may be banished from the pages of the journal.

Some of the pressure to publish comes, of course, from the need to demonstrate professional competence to administrators and others who control promotions, tenure, and related rewards. All those involved in judgments of scientists wish that there were some more objective and quantitative ways to evaluate the merits of published papers, and the degree to which they should be considered as positive factors in evaluating the author. Counting the number of references in a bibliography or weighing a pile of reprints is not enough.

I have developed a quantitative paper assay, a "Reward Index" (RI) system for use at evaluation time. The index is achieved by solution of an equation that includes weighted values for the nature of the journal, number of published pages, number of reprint requests received within the first thirty days after publication, number of citations in the work of others within one year after publication, favorable mention in reviews of status of knowledge, and number of requests to present seminars on the topic of the paper within one year following publication.

RIs can be calculated for each paper offered as support for a particular decision (such as promotion); the cumulative RI is a nice comfortable number, large or small, squatting solidly in the center of what is otherwise a very subjective procedure.

The strategist will quickly see any number of possible permutations and extrapolations of these concepts which surround the act of publication. They are attempts to confront very real problem areas—flexure points in scientific armor—and as such they should be scrutinized carefully.

SUMMARY

There are few areas in the conduct of science that are more important or more neglected by practitioners than formal publication of results and conclusions based on research accomplishments. Good scientific writing is the consequence of study and serious application, and it deserves more respect than it gets. A combination of good science and good presentation, an excellent published paper is a joy to read and a worthwhile objective for every scientist. The steps leading to the ability to produce such a masterpiece are many, and are often overlaid with interpersonal components such as relationships with co-authors, reviewers, and editors. Such components usually become apparent by trial and error, but are amenable to categorization and description, just as are the rules of grammar and composition. The totality of good written scientific presentation thus becomes a triumvirate of proper science, superb writing, and effective interactions with peers.

CHAPTER 2

THE SCIENTIST AS A PERFORMER
Presenting Scientific Papers

Ten principles to ensure the award of "worst paper of the meeting"; suggestions for better paper presentations; the enjoyment of oral presentations; humor, pro and con; discussion devices.

Oral presentation of scientific results before a group of peers is a necessary, useful, and productive component of existence in science. Why then is it so often done ineffectively or downright poorly? Some of the answers are fairly obvious: the scientist may not have much training in public speaking; the scientist has misjudged his or her audience; the scientist is presenting the paper only to ensure that his or her travel will be funded; the scientist has not taken the required time to prepare a good oral presentation; or the scientist has merely read a prepared text (a deadly business). The reality is often that the scientist simply hasn't bothered to prepare for the event, *thereby defying an entire set of game rules in an extremely critical arena—that of his or her assembled peers, where judgment is apt to be the strictest and occasionally the harshest.* This behavior seems absolutely inexcusable, but often stems from unawareness of the rules, or unwillingness to be bound by any reasonable set of guidelines.

HOW NOT TO PRESENT A SCIENTIFIC PAPER

Assuming that unawareness is part of the problem of so many deadly dull oral scientific presentations, especially at society meetings, I have drawn up a short list of admonitions and advice on how *not* to present a scientific paper (you will think of many more):

- Read it word for word from single-spaced typed pages— otherwise it won't be precise enough.
- Give all the minutiae of procedure in great and exhaustive detail—even if you run out of allotted time before you get to your results.
- Never time your presentation in advance or run through a practice session, because there is a certain magic that will make everything turn out all right.
- Fill up all your allotted time, and ignore cutoff signals from the chairperson. Keep talking until you are physically removed from the podium.
- If, by remote chance, there is time for discussion, assume that all questions and comments are designed to cut you down—so respond defensively.
- The proper time to think about visual aids is about three days before the meeting. Take some quickie photos of typed tables, or better still, draw some rough graphs on ground glass slides in your hotel room the night before the paper is to be given.
- To ensure minimum visibility of data, make extensive use of opaque projectors or viewgraphs, which adequately distort most graphic material and which usually project any tabular material dimly enough so that most of the audience can ignore it. Another useful device is to switch back and forth several times between 35 mm and opaque or viewgraph projectors—the confusion engendered will suitably distract most of the audience so your train of development (if any) will be totally lost to them.

- If you insist on preparing slides in advance, make sure that they are undecipherable. Tables should have tiny printing and at least thirty lines of numbers to ensure maximum invisibility of data—freeing you from the likelihood of challenge. Be sure to have tables photographed in dim light, slightly out of focus, so the entire mess is fuzzy. Then leave the slide on the screen for a very short period, to thwart the eager types with excellent vision who may be seated near the front of the meeting room.

    Another good device is to express amazement or dismay at the slides, claiming that this is the first time you have seen them. Periodically after that, apologize profusely and repeatedly for their poor quality, but make sure that they are downright awful.

    Still another ploy, if you have not prepared slides at all, is to claim that they have been lost or damaged in the mail, or promised by a junior author but not delivered. This gives you an opportunity to scribble illegibly on a blackboard if one is available in the meeting room, or to complain about its absence if one is not available.

- When you present the paper, act genuinely disinterested in the data and the audience; never, never give an indication of enthusiasm for your research; speak in a low monotone without pitch variation; and move away from the microphone whenever possible.

- If your paper is given in the late morning or afternoon, make sure that you start your presentation with a slide, and keep the room moderately to totally dark during the entire paper, with slides interspersed throughout. Many members of the audience will appreciate the opportunity to slip out of the room or to indulge in a short nap, and you will be reasonably free from unwelcome questions or comments when the last slide is shown and your paper ends simultaneously.

SUGGESTIONS FOR BETTER PAPER PRESENTATIONS

Verbal summarization of research findings before assembled colleagues is a vital aspect of a scientific career, and should be carried off with finesse. To do so is not a simple task; it requires hard work and careful planning, but is worth the effort—to all concerned. Much could be said about the processes involved in preparation and delivery of a good scientific paper; I have tried to distill my comments down to a few suggestions out of the many which might be proposed:

• The *structure* of the presentation must be given careful and extended thought. Emphasis must be on results and interpretations, rather than on techniques. The introductory few minutes could be well spent in placing the research in some historical and developmental context. Never underestimate the importance of an excellent introduction and a stunning conclusion—the critical first two minutes of your talk, and the final one minute.

• Practicing the presentation before a small group of supportive colleagues can do much to ensure an effective performance. Frank comments, and acceptance of such comments, are basic ingredients. Care should be given to correct timing, quality of slides, crispness of summary, style of delivery, elimination of annoying mannerisms, and supportability of conclusions. With this kind of preparation, a good paper is virtually assured.

• Good—even excellent—slides are rarely seen in most scientific meetings, and are warmly welcomed by the session participants. Production of such slides is not easy, and must be planned well in advance of the meeting date. Professional assistance in preparing visual material is well worth the cost, even if you must pay for it personally. There are so many simple devices—color, backlighting, sequential progression of data presentation, three-dimensional graphs, etc.—that are readily available, and that can transform a mediocre presentation into a memorable one. A good rule of thumb is to have 60% of your slides in color, and at least 40% of them legible from the back of the session room.

• As an important qualifier of the preceding, be very careful *not* to base a talk on successful slide presentation. Never begin your paper with a slide without establishing audience contact first. Slides must always be *adjunct* and not *integral* to the oral presentation, since dreadful things happen with great regularity to projection equipment. In more and more meetings we see growing abject dependence on visual aids, to the point where some scientists would be almost mute if they were denied a familiar carousel of slides or if the projector died.

Let me tell you the ultimate nightmare story about the folly of abject dependence on visuals as crutches in oral presentations. This one involved as principal actor a scientist who should have known better— a mid-level bureaucrat from an EPA regional office. He was on the late afternoon program of a scientific society meeting to talk about a favorite EPA topic—risk assessment—and undoubtedly had a carousel stuffed with beautiful, professionally prepared 35mm slides. Because of scheduling problems with another meeting, the conference room where the talks were to be presented had been changed; the newly assigned room was on the western side of the hotel, and <u>was equipped with blinds that could not be closed against the bright afternoon sun</u>— so the room couldn't be darkened—even a little. The bureaucrat had no option but to proceed. After being introduced by the session chairman, he gave a two sentence introduction (which apparently constituted the extent of his prepared written notes) and then called for his first slide. It was projected, and <u>it was virtually invisible, both to the audience and, more importantly, to the speaker, who obviously expected to read from it, and, sadly, seemed to have no written notes on the material it contained.</u> He was silent for a moment, and then *called for his second slide, which was equally invisible to the audience and to him—as was the third and fourth.*

After a long uneasy pause, with just a hint of embarrassed titters of laughter from the usual rowdy element in the back rows, <u>the speaker began his two sentence introduction again.</u> When it was over, he paused for a long forlorn minute while the bright afternoon sun slanted across the projection screen and the audience stirred uncomfortably. He then turned and walked silently out of the session room— never to return, not even to retrieve his unused slides. (I swear that this episode really happened, exactly as described here. I was part of the audience in that sunny conference room.)

If anything goes wrong with slides or projector, do not let bumblers or volunteers try to correct the problem. You will lose whatever attention from the audience that you have achieved, and most projection problems stretch on to ludicrous levels when inexperienced people try to solve them. By far the best tactic is simply to cut off any slides or reference to slides at that point, without apology or further comment. (It has occurred to me that a short evening or early-morning course in operating a 2 x 2 projector could be required of any person who might touch a projector during the meeting.)

The almost unbelievable extent to which things can go wrong was illustrated recently at an agency briefing meeting held on a New England university campus. This is an eyewitness account:

The culprit was, as usual, a 35-mm carousel projector. The first slide was glass-bound and predictably became wedged in the machine. There followed the usual outpouring of gratuitous amateur advice from members of the audience and a scurrying around for tweezers. The next slide was upside down. Then the machine went fully automatic and sped through the speaker's entire slide series in about thirty seconds. After this problem was corrected and the audience began to regain some decorum, the machine, of its own volition and without

human guidance, began advancing slides at odd moments, while pre-
vious slides were being discussed.

Fortunately the speaker was a veteran of previous warfare with
projectors, so at that point he walked off the podium, rapped the pro-
jector smartly with the wooden pointer, picked the machine up and
shook it briskly, pulled the plug, and resumed his talk without ever
again referring to slides. The whole episode was so hilarious and so
unbelievable that some thought it was staged—so many bad things
simply could not happen sequentially in one talk—but it was totally
impromptu. It did lead, though, to the thought that speakers might
consider using projection glitches as attention-getting or humor-
gaining devices. This approach must be well done if attempted at all,
but I can remember watching in fascination as slides made of non-
heat-resistant plastic gradually turned brown and their contained
data disappeared from the screen within a few seconds after the start
of projection. I can recall too the brief pleasure of a session with a pro-
jector which ejected slides so enthusiastically that they popped into
the air and into the audience—and no one seemed capable of correct-
ing the problem.

• An additional important but almost completely unen-
forceable rule in presenting slides is to *never let those responsible*
for session logistics plunge the meeting room into total darkness. Even
if subdued lighting detracts from your slides, *insist on that sub-*
dued lighting. This is almost a lost cause though; there is always
some eager person who will flip that final switch. Knowing that
this will happen, a good fallback position is to group slides into
small bunches; then when one set is finished you can call for
lights to revive the audience until the next set. Another, but
much less desirable approach is to save your slides until the
conclusion of the presentation, and use them as a summary.
Otherwise, if slides are interspersed throughout the talk, two
bad things happen: (1) you find yourself using slides as
crutches, your talk degenerates into a slide show, and you are
vulnerable to absolute destruction if the projector fails; (2) your

audience, in comfortably anonymous darkness, gradually nods off or slips quietly away from the session room to find other pursuits.

• Avoid at all costs the use of transparencies or opaque projectors. The state of the art for such equipment is primitive and invariably results in dim, fuzzy, and distorted material, especially in large rooms. People addicted to the use of such equipment also seem to be afflicted by another syndrome— they prepare transparencies of typed material which they then proceed to *read* word for word, or close to it. This is such an obvious crutch that it is beneath the expected competence level of any speaker, and is almost insulting to the meeting participants.

• In any presentation, assume that *whatever catastrophes can happen will happen*. This outlook makes it much easier to confront failure of the microphone, feedback sounds from the microphone, projector bulb failure, lodging of a slide part of the distance into the projection frame, workmen hammering behind a temporary partition in the session room, and sirens in the street immediately outside—all with a calm, unflappable external appearance, and with internal serenity as well. The alert game player will turn such events into assets by squeezing a little humor out of the situation. Responses to almost all events in the above categories can and should be *planned*.

Computerized projection is, of course, the newer approach to visuals for technical presentations, and the possibilities are limited only by individual imagination and the availability of time to invest in developing the necessary expertise. Some current applications can produce enhancement of lines or sections of text, progressive construction of increasingly complex diagrams on-screen, changes in perspectives in data presentations, zoom examination of geographic features, or graphics illustrating progressive seasonal changes in environmental variables. We are almost to that stage in visuals where, if we can think of it, it can be done.

But we still have not escaped from dependence on mechanical equipment—whether it be a transparency projector, a 35mm carousel projector, or a computer—for a successful presentation. The standard advice is, "Don't trust any piece of projection equipment to function properly, and never, never, prepare a presentation that depends in any significant way on visuals for its success." The failure of projectors is an endemic disease in every venue in which scientists gather, and the virus that has infected 35mm carousel projectors for decades has now been transferred successfully to computerized projectors. Even if the applications are simple—like the progression of enlarged succeeding section headings as a talk develops—equipment will fail, or will somehow be found to be incompatible with what is available.

The reality of equipment malfunction—even for the best state-of-the-art computers—was emphasized for me at an international meeting held recently in Baltimore. It happened that a German colleague who was also a personal friend had been invited to give the keynote presentation. He is an excellent speaker, even in his second language, and is known as an innovator in visual support for all his presentations. He arrived in Baltimore personally transporting the latest in German technology relevant to computerized projection. Since he was first on the morning program, he arrived in the giant auditorium fifteen minutes early to test—in typical German methodical fashion—his projection sequences. Signs of trouble appeared almost immediately: some minor incompatibility prevented the smooth flow of information. What survived were dim outlines of what should have been superb graphics, and no one had an immediate solution to the problem.

But the good Professor Doctor was a pro, and not the sort to be diminished by a reluctant machine. He had brought with him a backup set of standard 35mm slides, but they were in his hotel room six blocks away. No problem. In an instant he was down the escalator and into the street, running at top speed to his hotel. He arrived back in the auditorium with his backup slides just as the mayor of Baltimore and the Conference Chairperson were concluding their boiler-plate wel-

coming speeches. He went on stage as scheduled, with a truly brilliant presentation; few in the audience were aware of the tiny computer-related drama that had preceded it.

• Controlled but obvious enthusiasm for the research being reported is an element of great significance to a good presentation. This statement may seem almost platitudinous, yet it embodies a concept that seems to elude many scientists.

• Distracting mannerisms characterize the presentations of many speakers, and can often be eliminated by awareness and practice. How many of the following might be possibly associated with your oral presentation?

> Use of "ahs," "ers," and "ums" to fill gaps;
> Use of add-on nonsense expressions such as "OK?" (as a question), or "you know" (as a fill-in for slow mental processes);
> Swaying in and out of optimum microphone pickup range;
> Mumbling;
> Nervous scratching of body areas, even pelvic areas;
> Repeated clicking of a retractable pen;
> Rubbing hands together;
> Nervous manipulations of eyeglasses or jewelry; and
> Leaning on the podium, often for the entire performance.

• Effective presentation of scientific papers depends on a *conscious learning process.* Close observation of skilled presentations by others, careful planning of papers, confidence in the data, willingness to attempt new approaches, continuous practice in any kind of oral presentation, requests for critical comments on presentations—all contribute to a professional level of delivery of scientific products.

• Good public speaking should be part of the background and expertise of every scientist—but is often not. A surprising percentage of so-called professional scientists, even those from university faculties, are abominable public speakers. This is and should be intolerable to game players, who recognize that effective presentation of research findings is a critical part of the practice of science. Why, when so many avenues are available for self-improvement—Toastmasters Clubs, university extension courses, speech therapy courses, private tutors, even Dale Carnegie courses—must we be subjected to poor podium performances by presumed professionals? The answer is simple: We as listeners don't object vigorously enough to mediocre performances; we wait patiently until the amateur is through; we clap politely instead of booing; and we may even prostitute ourselves by trying to find something good to say later about a miserable performance. We therefore deserve all those deadly boring hours spent in darkened session rooms as captives of amateurs. Game players will do better.

• Beyond mere paper presentation at meetings, probably no single accomplishment is more important to success outside the laboratory than the ability to speak excellently in public. Years of training and practice are required, but the investment pays off beautifully in career advancement and in self-satisfaction. Scientists who are good public speakers are unnecessarily rare, despite the ready availability of self-improvement methods. Learning to be a good speaker is much like learning a foreign language well—prerequisites are minimum natural ability and maximum willpower and application.

The windows on a broader world that are open to good scientists who are also good speakers are almost limitless. Seminars, invited lectures, foreign symposium presentations, keynote speeches, banquet speeches, television appearances, visiting lectureships—all are available to the accomplished performer. Why, then, do so many settle for mediocrity?

• Scientific audiences are not basically different from those at theatrical or sports events. They have a fundamental interest in the event at hand, they have often made financial and other sacrifices to be present, and they expect a professional performance. *With these positive factors in operation, control of a scientific audience should not be complicated,* but it often is. A bumbling introduction by the chairperson, an uninspiring introduction to the paper by the author, a boring recitation of methodology, a collection of mediocre or poor visual aids—all serve to put distance between speaker and audience, and to ensure that members of an audience will not become *participants* in a scientific adventure.

Game players, with careful and thorough preparation in oral presentation, will seize control in the early seconds at the podium with a superb introduction; they will then move quickly to their significant findings and conclusions; and they will intersperse good oral presentation with exceptional visual aids and where appropriate a touch of humor. They will deliberately involve the audience in every step of their thought processes, and they will present conclusions in a way that invites comments and discussion. These are the things that transform scientific sessions into memorable occasions—these are the things that are the objectives of game players.

HOW TO ENJOY PRESENTING A SCIENTIFIC PAPER

If approached properly, presenting a scientific paper should be a *pleasurable* as well as profitable occasion for the speaker and the audience. The paper should be given in a way that appears relaxed and informed. Speakers often lose sight of the fact that scientific meetings are and should be *enjoyable* affairs—within the meeting room as well as outside. Speakers are a dominant part of the program, so if they are tense or stumbling or ineffective, then the listeners will be ill at ease and the pleasure of the meeting diminishes for all. Speakers have no right to do this to their peers.

On the side of increased pleasure for the speaker, there are a number of rules of behavior of scientific audiences that must be understood. A basic assumption is that probably only a handful of people will come specifically and purposely to hear any particular paper—or that if more come, only a handful will really give a damn about what is being said. The others may have wandered into the wrong room, they may be just waiting for the next paper, or they may have been standing too long in the halls (attending to the real business of the meeting, which is person-to-person communication) and need a place to rest. Audience numbers tend to peak just before the midmorning coffee break and again before midafternoon. Heights of peaks decline sharply after the first day of the meeting, as participants find more attractive pursuits outside the meeting hotel. Regardless of audience size, however, the speaker must take the situation lightly, and *must enjoy himself* or *herself*, especially since no one else is taking him or her, or the information, that seriously anyway. Good material will eventually appear in journals, and abstracts are usually available at the time of the meeting.

Some speakers, especially younger ones, tend to be unnecessarily upset when members of the audience (singly or in groups) suddenly and noticeably walk out of the meeting room halfway through the presentation. This is entirely normal—the rule rather than the exception—and usually has absolutely no connection with the quality of the presentation. Some reasons for such seemingly ungracious behavior are a necessary trip to the rest room, a need for fresh air, a previously planned but temporarily forgotten meeting with a colleague, an early plane, a planned shopping trip, or an overwhelming need for an early cocktail. None of these should constitute a personal affront to the speaker or the material.

The *use of humor* in scientific presentations deserves some attention, since it can be used effectively but also can be misused badly. Some reasonable guidelines are in order, although effective use of humor defies full encirclement:

- Introductory jokes unrelated to the subject of the paper should be carefully avoided, but ways can be found to insert genuinely clever, unobtrusive, and *uncontrived* humor. This can be particularly effective in the early moments of a paper, to both alert and relax the audience.
- Light remarks are in order when the projector bulb blows or a slide appears upside down—but don't attempt humor at the expense of the projectionist, since he may be a graduate student recruited at the last minute and unfamiliar with the equipment, or worse still a union man who could go on strike immediately if his feelings are injured.
- Exceptional and humorous slides can be prepared, but they must be very carefully considered lest some in the audience feel patronized or otherwise insulted. A *good* cartoon, relevant to the moment, may be effective, whereas several consecutive ones might be completely out of place. Good candid shots can occasionally add a light moment.
- Attempts at humor in oral presentations should usually appear to be spontaneous, or should be very clever without being "cute." Humor is of course an art form, and as such it must be handled delicately and with care—otherwise it should be strictly avoided.

There are several good methods of ensuring that an oral presentation will be pleasurable for the speaker as well as for the audience:

- The speaker should have made a serious attempt to evaluate the audience, and to package the talk to fit it. This is not a simple process, since decisions about vocabulary, style, and approaches must be made in advance, from a distance.
- The speaker should have great familiarity with the subject matter to be included in the paper. (This would seem

almost axiomatic, but it is surprising how many speakers seem unsure of their material.)

- The speaker should walk to the podium with the thought that he or she is among friends who are interested in the material, and with the conviction that what will be said is meaningful and important.
- The speaker should single out a few responsive faces in the audience (they occur, fortunately, in every group) and concentrate on them in sequence.
- The speaker should give nonverbal as well as verbal cues that the occasion is a pleasant one. An appropriate smile or two, good audience contact, a preplanned and relevant light remark—all of these can convey the positive feelings that the audience has a right to expect.

THE SO-CALLED DISCUSSION FOLLOWING PAPER PRESENTATION

The paper may have been presented beautifully, with clear sequences of data, good development of conclusions, good diction, a few good slides, and significant findings—but occasionally the discussion which follows is a disaster. Explanations may be inadequate, admissions about unfamiliarity with certain literature may be made, untenable generalizations may be refuted—in short a whole series of pitfalls can await the unwary during that critical pregnant period which follows the last ringing sentence of the formal presentation.

It is here, in this crucial round, that the good game player can emerge as an obvious winner—assuming of course that knowledge of the subject and the data are reasonably sound. There is no reason to flounder around, or to apologize, or to admit ignorance too easily.

Those of us who have attended enough scientific meetings have witnessed "indecent exposure" situations where speakers have been sliced and shredded badly, often because they were inexperienced or unprepared. Such episodes are a reflection of unfamiliarity with game rules on the part of the shredder and the shreddee. The shredder—often a senior person experienced in postpresentation repartee—chooses to ignore the rule of gentle and painless surgery in favor of gross mutilation. Sometimes this is warranted, if the paper and/or the data are poor and deserve negative comment, but sometimes the action is gratuitous and could be done more gently, possibly in private conversation with the speaker. In most societies there are sadistic individuals, often capable scientists, who are noted as seeming to derive great satisfaction from what appear to others to be unnecessarily harsh comments. (It may be of course that such individuals have their own game rules, which are different from ours.)

At any rate, the shreddee is largely responsible for his or her own survival and well-being in the discussion period. There are many ways to avoid entrapment by the few individuals described above, and to make the discussion a mutually profitable experience for those who truly want to explore the data and conclusions presented. Some of the best (from the speaker's point of view) are to know the material well, to discuss any weaknesses in the data with sympathetic colleagues in advance, to avoid being overly defensive about conclusions, and even to write out detailed answers to a number of possible questions that might be asked. Bluffing rarely succeeds, but abject confessions of ignorance are not required either. Questions can often be responded to truthfully if only partially (a lesson learned early by politicians). Responses to questions can also be diverted into areas more familiar to the speaker, as long as the diversion is not too obvious (another lesson politicians learn early).

Fortunately there are few absolutes in science, so incompleteness of data and/or inconclusiveness of findings are not fatal flaws, if presented in a positive, undogmatic manner. As in so many interpersonal exchanges, success in handling discussion questions and comments effectively is largely a matter of practice, building on a few essential game rules and on a core of personal experience and reading in the subject matter area.

THE LAST WORD

Presenting a scientific paper before a gathering of peers can be a valuable and pleasurable experience for all concerned. The speaker is obviously one of the principal actors in what is for many societies an annual event of some significance. As such, he or she owes the audience at least the following: obvious enthusiasm for the subject matter, emphasis on significant conclusions, avoidance of exhaustive details of technique, effectiveness and conservatism in use of visual aids, and reservation of adequate time for discussion at the end of the presentation.

Scientific audiences play a vital role in scrutinizing data and conclusions given in oral presentations, and in pointing out deficiencies and inadequacies prior to publication. The entire interplay can be structured to provide a maximum of profit and enjoyment, and a minimum of pain and discomfort, if participants know and follow a common set of rules.

CHAPTER 3

THE SCIENTIST AS A FACE
IN THE CROWD

Attending Scientific Meetings

The scientific meeting as a social experience; the ins and outs of session attendance; activities outside the session chamber; the significance of cocktail parties; election to society offices; breaking into the international circuit; was it all worth it?

We considered in the previous chapter some aspects of presenting papers at scientific meetings—a very worthwhile activity for the few. But what of the many others who regularly or irregularly attend society meetings or symposia but do not feel compelled to present papers? Why are they there? How do they get funds to attend? What do they gain from the meeting? Answers to these and related questions constitute the framework and substance of this chapter.

STRATEGIES FOR MEETING PARTICIPATION

We have found through several statistically unreliable surveys conducted at different kinds of scientific meetings—

from the small regional chapter meetings of national societies to large international symposia—that most attendees claim to be there primarily to keep up to date on the latest developments in their fields. This is a laudable but highly suspect response, when so many journals can help them do just that. A more plausible response is that meeting attendance enables *personal interaction* with friends, peers, and authorities. *The meeting, then, should be considered as primarily a social event—a pleasant interlude with colleagues, away from the laboratory and classroom.* It is a temporary assemblage of people with reasonably compatible views of the world, and as such it provides opportunities for personal contacts, field trips, expeditions to local restaurants and nightclubs, and late-evening discussions in hotel rooms. These are the ingredients and events that will be remembered, rather than the scientific papers that are heard or presented. Once this fundamental insight is perceived and accepted, then all the various components of the meeting—the scientific sessions, the mixers, the committee meetings, the banquet—assume their proper dimensions within the total montage of the meeting.

The foundation must always be the scientific sessions of contributed papers, but those who equate the totality of such sessions with "the meeting" are usually relative newcomers and definitely not game players. There are so many other good, useful, and productive aspects of scientific meetings that it would be a minor crime to conclude that the sessions are all-important. This brings us to Fundamental Rule #1 in meeting attendance:

> <u>Plan to attend only very carefully preselected</u> <u>papers, of the many that are offered. Never, never sit</u> <u>through an entire session of contributed papers,</u> <u>since most of them will be irrelevant to your inter-</u> <u>ests</u> (and you can read the abstracts at your leisure, if necessary).

The papers you select might include those of your friends, students, or close associates; those that offer some overview or new approach; or those given by especially good people in the field (unfortunately some of these can prove to be a waste of time too).

You should never feel trapped into sitting through an entire sequence of papers unless you wish to, or unless they are part of your preselected list. You should not feel the slightest restraint against quietly rising to your feet and departing from the meeting room at the end of any paper. You should, however, feel some proper mild embarrassment about lurching to your feet and leaving in the middle of a paper (especially if the lights are on). Even though the speaker, if he or she is a game player, will not be flustered by your exit, it seems only fair, as a general rule of thumb, that if you stay until the speaker begins you should feel committed to stay for the entire paper, as a matter of common courtesy if not of interest.

Assuming that you do manage to sit through an entire paper, the discussion following each presentation affords an opportunity for questions or comments relating to clarification or expansion of points made, but can be used for other purposes as well. Some game rules for discussion participants could be cited, but they would be ignored by at least two categories of perennial meeting-goers: (1) The "compulsive discussant," who feels some fiendish urge to make comments after every single paper. This can be okay if the audience is sluggish anyway, but for an active group it can reduce the time available for other participants to ask questions. (2) The "informal paper presenter," who abuses the discussion period to present an informal paper (often complete with slides) relating to the one just given. An experienced session chairperson can subdue such characters, but they may overpower an inexperienced one (see Chapter 4 for control measures).

Lessons to be learned about discussion periods include such things as the proper way to ask questions of speakers who are authorities (answer: "Carefully, so they don't launch into another ten-minute speech") or speakers who are graduate students (answer: "Reasonably gently, but don't let them get away with anything smug or dogmatic or pedantic").

In addition to the technical sessions, many meetings include keynote speeches, invited lectures, special symposia, and panel sessions—which brings us to Fundamental Rule #2:

> Plan to attend all keynote speeches, invited lectures, special symposia, and panel sessions in your area of interest (or even in peripheral areas).

Almost invariably the people selected for these special events are highly qualified as speakers or as experts (and often both). Such events therefore constitute an opportunity to gain a painless overview, and also to study techniques of presentation—things the game player is always alert to acquire. Furthermore, attendance at these events should be considered part of your commitment to the society and the meeting. The organizing committee for the meeting gets little praise anyway, and the very least that registrants can do is to contribute their physical presence to the special events.

A logical question at this point might well be "If I attend the events and only a small selection of technical papers, where do I spend the rest of my time (calculated to average about 60 percent of the time between 9:00 A.M. and 5:00 P.M.)?" This somewhat naïve question leads to Fundamental Rule #3:

> The purpose of meeting attendance is to see and talk to people, and to discuss your specialty with colleagues, outside the session chamber.

Remember that you have not come to the meeting to be stuffed like a sausage skin with new scientific information. Rather, you have come to communicate with peers, to hear about the latest moves and victories of close colleagues, to discuss the grant situation, and even to look at new publications and equipment—all at your own pace. Once this concept is fully accepted (and it takes practice to avoid drifting back to the session room, even if the presentations do not interest you), then it is possible to really enjoy the meeting.

Of course there are traditional formal social activities usually associated with most meetings—luncheons, banquets, and mixers (variously called cocktail parties, cash bars, happy hours, attitude adjustment sessions, etc.). Fundamental Rule #4 applies to such structured affairs:

> Luncheons and banquets are usually lost causes unless you pick your tablemates carefully, but the cocktail parties or mixers deserve your whole-hearted and enthusiastic participation, since this is where much of the business of the meeting occurs.

A basic operating principle at cocktail parties associated with meetings is a simple one—get there very early, as soon as the bar is open, before the hors d'oeuvres have disappeared and before the noise level has reached a point where conversation is impossible. Another operating principle is that if you want to discuss a point with people there, especially the "great ones," do it early, before the third drink, and before disciples and hangers-on form tight and almost impenetrable circles around their mentors. After this early business period, you are then free to mingle with good and knowledgeable companions and friends for a combination of news gathering and shop talk. The best part of this early and late division of effort is that you can then select good companions for dinner or whatever local excitement the meeting city offers.

The operating principles are somewhat different for graduate students and junior faculty members in need of a job change. The cocktail party can be a very important encounter zone with the "right people," if you know someone to introduce you and if you are capable of making a good impression in such situations. Many people may not remember your name, but they may remember that they did talk to you and that you may or may not have said something brilliant (or at least relevant). The most brilliant thing that you can say would be some complimentary remark about the "right people's" research, and how his or her work contributed to your interest in the field.

There may be a number of wives of participants present at social functions, since more and more meeting organizers are sensibly including programs for spouses (assuming for the purposes of discussion here that the scientist is a male). Quite often the wives are brighter and more fun to talk to than their scientist husbands, but they get turned off quickly by shop talk unless they are scientists themselves. They are usually well informed about many things, and conversation with them can be a welcome change from too-intensive technical talk as the evening wears on.

Surveying the total meeting experience, *participants should remember that a committee of their colleagues has spent a significant number of hours away from research and/or teaching to organize the meeting.* Regardless of the perceived inadequacies in planning, the shortcomings in equipment and facilities, and the minor inconveniences that may emerge—no public expression of dismay or complaint during the meeting is warranted, unless it is about something easily corrected by the organizing or program committee.

As an extreme example of poor taste, with negative impact on a harassed local organizing committee, I cite an episode that occurred at an international meeting in the Caribbean several years ago. A speaker actually took it upon himself to spend the first few minutes of the time allotted to him for scientific presentation to complain about the meeting room facilities (lighting, microphone, projection, even the air conditioning), the session organization (too little time for his presentation in particular), and even the meeting location (in a delightful if somewhat remote area).

Now admittedly this is an extreme; the more common outlet for minor frustrations would be comments to other participants during coffee breaks, or possibly a brief expression of unhappiness to the committee chairperson about some deficiency in the program or facilities—often thinly veiled as humor but usually not at all humorous to the organizing committee. A good rule here is one of public silence: resist any urge to criticize efforts of the unpaid committee. Treat observed inadequacies or failures as learning experiences, not to be repeated when your turn comes to organize.

If, however, the inadequacies result from inaction or deficiency on the part of a paid executive secretary, a paid meeting organizing company, or the staff of the meeting hotel, then the guidelines are quite different. A request should be made for immediate correction if it is feasible, and there is every reason for objective correspondence after the meeting, preferably with the president of the society, offering criticisms and suggesting remedies for the future.

I encountered a good example of a problem and its solution at a recent society meeting held in one of the new high-rise waterfront hotels in a major coastal city. The hotel reception people were downright "snotty," session rooms were not prepared, and a number of confirmed reservations were not honored. It was obvious that the meeting organizing company, which had been paid to handle all the nonprogram aspects of the meeting, had not performed adequately and was not about to act. Representations by a number of participants, through the president of the society to the hotel management, brought significant changes almost immediately, and subsequent pressures will undoubtedly result in a different choice of organizing company for future meetings of the society.

But these minor inadequacies are exceptions. The bottom line here is, as always, that scientific meetings should be *pleasant* as well as *profitable* experiences. Successful meetings are almost always by definition pleasant; those actions and events which contribute to pleasure and profit for all participants should be encouraged, and sour notes should be avoided or minimized wherever possible.

FUNDING MEETING ATTENDANCE

Finally, after the last paper is presented and the last cocktail consumed, it is logical to consider a question that probably should have been answered at the beginning—*"How do I get funds to attend the delightful social event called a scientific meeting?"* First of all, you never admit to those who control the funds that the meeting is really anything more than round-the-clock scientific presentations that will benefit you enormously in your job performance. Having not made that admission, the most important positive step is to begin advance planning long before the meeting (up to a year, especially for international meetings). Advance planning might include early contact with the program chairperson or the organizing committee, indicating interest in presenting a paper or suggesting a good concept for a panel discussion or a workshop that could form part of the meeting. Often a positive response from the meeting organizers can be parlayed into active participation in the program and approval of travel funds from appropriate administrators.

Another approach, if you happen to know any of the session chairpersons, and you are qualified, is to ask them flat-out for an invitation to serve as speaker, discussant, or panel member, depending on the structure of the session. Such an invitation can then be placed gently under the nose of an administrator who controls travel funds.

Still another effective approach is to have some contact with a laboratory or university in the vicinity of the meeting. If you know any staff or faculty members there, an invitation to present a seminar at the approximate time of the meeting provides a double reason for approval of funds for travel.

If all these strategies fail, and if no grant money is available for travel, you could contemplate the almost unthinkable—paying for the trip yourself and treating the whole thing as a vacation, since scientific meetings are usually such pleasurable affairs. (One small added compensation here is that a limited amount of legitimate professional travel may be tax-deductible.)

ELECTION TO SOCIETY OFFICES

Scientific societies are in some ways like social clubs; their members have common interests and enjoy one another's company, even though the meetings may occur infrequently. Society affairs are guided by a constantly changing cast of unpaid officers, elected for one- or two-year terms. The short tenure gives many members, successively, the opportunity to make some impact on the course of events, with only minimal diversion from other professional activities. Election to society office thus becomes an attainable goal and another pleasant aspect of a scientific career.

A little planning is in order if society activities—including the transient honor of office holding—are to be enjoyed to their fullest extent. Some guidelines for upwardly mobile junior scientists include the following:

- It is expedient, early in a scientific career, to focus on a few societies which are most concerned with your area of interest and expertise. A larger number can be joined, at least temporarily, but remember that it is difficult to participate actively in too many societies, since each involves a reasonable time commitment and this is time spent away from research and/or teaching.
- It is expedient to join some local or regional societies, as well as the big, impersonal national ones. Each type will provide for particular needs—the big ones have prestige

and national membership; the smaller ones offer greater opportunities for continuing personal contacts.

- Present good, even exceptional papers at meetings of the societies.
- Attend society business meetings and have meaningful comments to make.
- Talk to people in the corridors about *their* research, and about *yours* only on request.
- Introduce yourself whenever it is possible and non-intrusive, but have something sensible to say after you've said hello.
- Don't depend on your conference badge for identification. Most of them are typed with tiny letters, and some people's eyesight isn't that great. Always introduce yourself with name and affiliation.
- If you are a graduate student, go to meetings with your major professor, provided that he or she is willing. Appear with your professor (with discretion and at his or her invitation) at cocktail parties and in the corridors, and get introduced to some of his or her colleagues in this way. (Don't make a nuisance of yourself, however, and do have something relevant to contribute to the conversation.)
- If you hear of open late-evening cocktail parties in hotel suites—*go*—even if you hate alcohol. You'll meet a mixed bag of people, some of whom will be interesting and some of whom may even remember your face if not your name the next morning.
- In every society there are several categories of "in groups." Examples are officers, boards of directors, authorities in particular specialities and their sycophants, nightclub circuit riders, porno movie buffs, and restaurant connoisseurs. Join one or more of these if access is provided and if you're interested—but don't intrude otherwise.
- Volunteer, without being "pushy," for any entry-level society job that appears—especially if it gives you some

visibility. Most societies have numerous committees, and each chairperson is alert for bright junior people who are willing to work.

- If you get a committee assignment, do about 200 percent of what is expected.
- Election to the first society office is a critical step. One direct approach is to learn the names of the members of the nominating committee and to cultivate them—carefully and unobtrusively. They may remember you at some point in their deliberations. If you are nominated, keep in mind that campaigning is usually not overt, but campaigning is done, very discreetly. Don't push too hard or too obviously for any position in a scientific society, though, since it is easy to turn people off if they don't know you well.
- If elected, try to spend some time with the previous incumbent. There is much to learn about people interactions, whether the job is board member, secretary, treasurer, or other entry-level position. Much of this information is available free from your predecessor, if you approach him or her correctly.
- Once elected, do 200 percent of what is expected in the position, and do it with enthusiasm, perception, and grace.
- Most of all, enjoy society meetings, either as a member or as an officer. They can be sources of some of the lasting satisfactions and pleasures of a scientific career. Some of your best friends may turn out to be scientists.

INTERNATIONAL MEETINGS

National society meetings and various kinds of national workshops constitute the "journeyman level" of scientific meetings, but the peaks are clearly located in the international symposia or conferences. Most of them are convened irregularly, and may be one-time events; some recur every year, every other year, or on some other regular schedule. They may be sponsored by international organizations such as the United Nations, or by international federations of scientific societies. Most of the presentations are by invitation, although limited numbers of contributed papers are sometimes accepted.

Participating in international meetings can be an enlightening and valuable experience in many ways. It is possible to examine the stage of development of particular scientific disciplines in a number of countries; it is possible to visit foreign laboratories and field installations; and it is possible to establish communication (and even cooperative programs) with foreign scientists. Often such contacts can lead to extended foreign visits for scientific purposes, under a variety of existing scientist exchange programs.

A logical early question is "How do I become involved in international science?" The answer is, as you might expect, complex and unsatisfying. First of all, you must have scientific credibility and expertise as export commodities (this is not necessarily so for science administrators, who may have other export items, such as dollars). Beyond that, you need some existing communication with foreign scientists in your specialty. This part is easy; many of us correspond regularly with a number of foreign scientists, if merely to request reprints or comment on their published work. Correspondence does not necessarily have to be with the recognized authority in a particular country (but it helps). There are many good scientists in countries such as Japan and Russia, most of whom do not travel extensively, and most of whom publish in languages other than English. Fruitful, mutually advantageous contacts may be developed with such people simply by writing (not always so simply in the case of Russia, in my experience, but this is unusual in international science). It is highly advantageous to have a good working knowledge of at least one foreign language, beyond the level required to pass Ph.D. qualifying exams.

The first international scientific conference or symposium can be important, provided enough advance planning has been done. You will have initiated correspondence with several counterpart scientists in the country where the meeting is to be held, and in the second or third letter you should indicate the possibility of your attending the meeting and ask if there would be a chance to visit their laboratories. You should brush up on or take an intensive short-term course in the language of the country in which the meeting is to be held (even if simultaneous translation into English is to be provided at the meeting), and you should be prepared to use that language in conversations with foreign colleagues if you have achieved any level of proficiency. You should find out who the organizers of the meeting are and inquire about contributed papers. If such papers are permitted, send a substantial abstract, and have something meaningful, unique, and substantive to report. If no contributed papers are permitted, go to the meeting anyway, at your own expense if necessary, and contribute to the discussions where you can do so intelligently. The visibility of paper presentation is important, but the contacts that can be made, regardless of the paper, make the trip worthwhile.

There are of course many other routes to participation in international science; some of them can be planned, but many are entirely fortuitous. There are, for example, numerous bilateral and multilateral scientific panels, joint committees, and commissions. Many of them hold special meetings or conferences to which they invite national experts. Many government agencies participate in international discussions on mutual problems, and again they frequently invite national experts.

The United States is party to a great number of international agreements and compacts. Delegates to the regular meetings associated with such agreements (who are usually not scientists) like to have a cluster of national experts behind them as "scientific advisors." This is a distinct and demanding kind of international science that deserves brief mention. Scientific advisors are in attendance to provide authoritative detailed information to the delegates on extremely short notice. Often there is only one expert present to represent each area of science involved in the discussions; that expert will be expected to be available at all times during the sessions and to have answers *then*, not later. It is an all-or-nothing arena, suited only to those with a high level of expertise, instant and total recall, and a keen analytical mind—but it can be an extremely exhilarating environment for the qualified. Some important ground rules for scientific advisors include the following:

- Keep a low profile.
- Remember that the meeting is *political* and not scientific; decisions will consequently be based on political considerations and not necessarily on scientific rationality.
- Keep the official delegate(s) informed—insofar as information is wanted. This responsibility includes gleanings from corridor conversations, if relevant.
- Attempt to speak a foreign language, or if you absolutely must, speak softly and distinctly in unaccented English.
- Bring key reference material with you, but don't count on it to help you provide the kind of instantaneous responses that will be expected of you.

There is an added advantage to this kind of meeting in that delegates from the other participating countries also bring their scientific advisors, and outside the formal sessions there is ample opportunity to mingle and to discuss scientific matters (assuming language compatibility, which to an American usually means "can he or she speak English?"). Some of these scientific counterparts can make good dining and drinking companions. Often such contacts lead to future international scientific conferences or workshop plans and proposals, beyond the intensely political climate of the moment—in fact some of the best international conferences, in my experience, have developed from problems that surfaced at international commission meetings.

There is another, gentler, kind of international science as exemplified by meetings of such organizations as the World

Health Organization (U.N.) and the International Council for the Exploration of the Sea. Participation in the activities of these and similar groups can come through such routes as submission of research proposals with international cooperative activity at their cores, or by being invited to give a specialized paper in your area of competence. Often an initial involvement, if carried out exceptionally well, can lead to continued association with the organization. The United Nations, with its numerous scientific bodies and with projects all over the world, is an excellent organization for the development of international contacts, but it is often difficult to find out about what is going on, except almost by chance or by overhearing a corridor conversation among those actively involved and already "inside."

This section on international science would not be complete without some brief description of a category that I label "*world-class scientists*"—even though "world class" has become an overused buzz word. These people exemplify the best that each country has to offer, and they tend to gather at major international symposia and conferences, often by invitation of the organizers. Just as is the case with world-class athletes, it is a pleasure to observe the performances of such scientists—at the podium, in small-group conferences, in committee and working-group meetings, in summary sessions, and in informal groups. They differ from one another physically, but they are uniformly astute, urbane, perceptive, interesting professionals who have survived a severe selective process within their own countries. Most of these world-class professionals are consummate game players: they are politically aware, they are diplomatic by instinct and training, they are sensitive to nuances of interpersonal relationships, and they are often superb scientists. Furthermore, they are frequently found in leadership roles where science interfaces with political processes in many and varied international panels, commissions, councils, and advisory boards—often as a consequence of counterpart roles that they play within their own countries. Careful study of these world-class scientists should be made whenever the opportunity exists to observe them in the native habitat of the international meeting; this can be an unparalleled learning experience for the developing game player.

RESIDUES OF SCIENTIFIC MEETINGS

In examining various kinds of scientific meetings—annual society meetings, workshops or small-group conferences, and international symposia—it is logical to wonder about the benefits and the products of these often elaborate and expensive gatherings. The sound and fury normally lasts for no longer than a week, then the meeting dissolves and recedes into history and the participants melt away, leaving only the published proceedings as a tangible legacy. The excitement, the personal interaction, the charged atmosphere of good scientific exchange—all evaporate and are lost forever, except in the dimming memory of the participants. What, then, is the value of it all?

The saving feature of society meetings is that many of the papers will be published subsequently in the journal of that society, and the comments (if any) received on the oral presentation may help to sharpen and improve the written paper. Additionally, information is presented about the current status of research in many subdisciplines.

Workshops are easier to justify, since a summary report is and should be one of the expected tangible products, to be reworked by correspondence into a publishable document. Here, though, there are risks. The discussion leader may draft *his* or *her* report as *he* or *she* saw the meeting (which may not be as the participants did) or the rapporteur may have misconstrued some or much of what was said. A rational way to reduce misunderstanding and to reach consensus is to insist that all participants review the draft report, and that their comments be incorporated before general release.

Symposia often leave a legacy of the printed proceedings,

which may contain only the principal invited papers, or in some instances all the individual contributions. Only rarely in such large groups is any consensus reached about anything substantive. Opportunities for joint public statements are rarely taken, often because participants fear the "politicizing" of scientific meetings by such statements. There are notable exceptions. In my experience, a U.N.-sponsored world conference on aquaculture resulted in an elaborate "declaration" about the future of food production from aquatic sources—a statement that has been widely quoted and used as a basis for national action. A few year later, an international symposium on marine pollution produced a "memorandum" developed during the meeting which summarized the participants' concern about the extent of degradation of the world's oceans, and recommended action to reduce the impacts. The memorandum was widely publicized, particularly in newspapers of several participating countries. It is unfortunate that these are rare events; we should do more to distill the best thoughts and opinions expressed during such symposia into meaningful summaries or public statements which can be incorporated into the printed proceedings or released through the news media. Those responsible for organizing meetings of any kind should consider the broader implications of what is said and discussed—and should encourage public dissemination of the quintessential products.

Scientists, of course, do not want to be "used" by individuals riding particular hobby horses or espousing one-dimensional causes. Furthermore, most scientists resist using credibility gained in their own discipline to support causes and objectives unrelated to that discipline. However, a moment comes for some when they wish to be part of a public movement as individuals, whether or not the movement is directly related to their expertise. Thus biologists may speak out as citizens against nuclear proliferation, chemists may express individual concerns about genetic engineering, and physicists may object as individuals to abortion legislation, to cite just a few examples. This activity should not, however, be part of a scientific meeting.

SUMMARY

Scientific meetings are and should be social as well as technical happenings, characterized by peaks of interpersonal activities. Whether regional, national, or international, they provide forums for communication and interaction that are not easily available otherwise. This is particularly true of international meetings, where the best national scientists are often in attendance.

Most meetings have as their rationale and purpose the exchange of scientific information, but in addition participants may have many less-visible objectives—ranging the full distance from finding a new job to accumulating peers to help write a book. Much of the substance and pleasures of the meeting are to be found outside the session rooms—in corridor discussions, cocktail party shouting sessions, and late-evening small-group conferences.

CHAPTER 4

THE SCIENTIST AS A CONCERTMASTER

Chairing Scientific Sessions

Early planning—success or failure of the session happens here; the event itself, with pitfalls; the follow-up; special assignments.

There are numerous printed guidelines for the preparation and oral presentation of scientific papers, and almost every scientific society has felt called upon to produce one, but detailed information is not always as readily available for the guidance of session chairpersons. Since the session chairperson can make an important contribution to the success of a meeting, it seems worthwhile to assemble some possibly helpful suggestions, even though many of us may have already developed our own concepts of the ideal session chairperson through long (and sometimes painful) experience as participants in scientific meetings. Occasionally it seems that session chairpersons consider the invitation to act in that capacity as honorary, without much more responsibility than simply appearing at the stated time prepared to introduce participants as indicated in the printed program. A more dynamic approach to the job and active participation in planning the session would do much to enhance the value and success of a meeting, and is one more small ingredient in successful game playing in science.

This chapter is drastically different from all the others in this book. It seemed that somewhere there should be as complete and exhaustive a treatment of the "how to" aspects of a topic as could be produced. This is it. The other chapters can be described as "samplers," but here in this one is most of what I have learned about chairing scientific sessions. There is more here than can be used readily, unless chairing sessions becomes a full-time occupation or an obsession—and some of the material could be described as a matter of opinion, subject to argument.

Part of the reason for such an exhaustive treatment is that I have seen the job of chairing sessions done so poorly, so often, and so needlessly. Next to superb presentations of papers, there is little that contributes more to the success of a session than an outstanding performance by the person chairing it. In an earlier chapter I emphasized the *pleasure* that should be derived from attending scientific meetings. Sessions which are chaired smoothly and professionally contribute substantially to that pleasure, and are the responsibility of those selected by the meeting organizers to lead. We have all witnessed, at one time or another, but not consistently enough, the brilliant performances of excellent session chairpersons—good control of speakers and audience, wit, humor, sparkling and informative introductions, good discussions, relevant and thoughtful summaries—so we ask "Why isn't this the *rule* rather than the exception?"

The session chairperson's role should be played in three acts: (1) premeeting planning and communications, (2) guiding the session itself, and (3) postmeeting communication. In each phase, specific activities by the chairperson can help to assure a smooth, relaxed, and productive session. Since the structure and purpose of scientific meetings vary significantly, no single complete set of guidelines can be appropriate for every session—but in modified form, some general procedures will fit most occasions.

EARLY PLANNING

Once you accept an invitation to chair a session, you are immediately responsible for its success. You may choose to do a haphazard last-minute job, or you may invest enough time and thought in the session to make it genuinely outstanding. Much of your success will depend on what is done *before* the meeting, rather than during the meeting—although the latter is important too. The following suggestions about session planning should contribute to the value of the meeting and the success of the chairperson in that job.

• *One of the most crucial elements in planning a scientific session is the selection of participants.* Many times such selection is beyond the control of the session chairperson, particularly when sessions are based on contributed papers at meetings of large national societies. Often, however, you as the session chairperson are expected to suggest, recommend, select, and invite participants and discussants for colloquia, symposia, or special interest panels. It is here that you can exert maximum impact by choosing knowledgeable, articulate, and conscientious participants. You can blend youth with age, you can explore different approaches to a common problem, or you can provide multiple illustrations of a central theme. Speakers selected should be interested in effective oral presentation of their material, and you should place such an interest very high on your list of criteria used in choosing participants. Too often scientific audiences have been subjected to ill-prepared, uninspiring oral presentations of subject matter that may have been of importance or interest to the audience, but apparently not to the speaker. His or her manner of presenting the material may have left the group with the feeling that more might have been gained from a photocopied handout and ten minutes of silence. You as chairperson should assiduously avoid scientists known for this type of presentation, whenever possible.

• *Selection of participants and invitations to participate should be made as far in advance of the meeting as possible.* A definite positive correlation can be shown between percentage of acceptances and chronological distance from the meeting date. The correlation should be heeded by every session chairperson, but frequently it is not. Often initial contacts and tentative acceptances can be completed by phone and followed by a confirmatory letter. You may find that this device is particularly effective with the imposing number of scientists who are remarkably poor correspondents, and who may delay responding to your initial written invitation until you have given up and approached someone else.

• Once the formalities of invitations and acceptances have been completed, *each speaker should be informed in writing, well in advance of the meeting,* about the general structure of the session—its purpose or theme, who the other participants will be, and where each fits into the session. Often this can be done by several informative circular letters from you, as chairperson, beginning about six months in advance of the meeting date. Attempts such as these to reduce the uncertainties in participants' minds will usually produce a more cohesive and smoothly functioning session.

• *Speakers should be told the exact composition of their audience,* and should be strongly encouraged to pitch their talks accordingly. Speakers should be clearly warned *not* to prepare a two-hour talk and hope to compress it into fifteen minutes. *They should limit their material to fit the time allowed.* There is rarely room, even in the most specialized of sessions, for papers which include many detailed tables, numerous graphs, or lengthy explanations of procedures. Similarly, papers which are no more than slide shows should be discouraged for scientific meetings. Visual aids must be adapted carefully to the size of the audience and the session room. Unless the session is very small and highly informal, the use of blackboards, flip charts, or opaque projectors should be strongly discouraged—in fact you might go so far as to instruct participants that only unbound 2×2 slides will be accepted—absolutely no others, no large transparencies, no opaque projector, no power point projection, and no film-strips. It takes only a little advance planning to prepare slides of suitable material for visual presentation.

• *Each speaker must be informed early and specifically about any printed or typed material expected.* He or she must be told whether the completed final draft of the paper is to be turned in at the time of the meeting, whether an abstract is required (and when), and whether photocopies of the paper or summaries are expected (and if so, how many copies are needed).

• *If discussants are selected, they must also be intimately involved in presession communication.* You as chairperson should inform them well in advance of the meeting about their exact roles—how long a discussion to plan, whether it is to be a critique or extension of the paper presented, and other relevant details. The discussant *absolutely must* have a complete copy of the paper(s) in his hands at least two weeks before the meeting.

• You should request from each speaker (and discussant) well in advance of the meeting a *brief introductory biographical paragraph* including details the speaker would like to have included. At the time of the meeting you should, in your introduction, pronounce clearly and distinctly the full name, title, and affiliation of each speaker and the title of the paper.

• A small element of advance information which speakers appreciate concerns registration. Some meetings provide complimentary registration to session chairpersons and invited speakers (and sometimes their spouses); other do not. Some even provide travel expenses and/or honoraria. You should convey this type of specific information to all session participants well before the meeting.

THE MEETING

Careful planning is important to a successful session, but often after the most conscientious preparation, one or more seemingly unimportant details can, if overlooked, detract from the effectiveness of the session. Sometimes such details are beyond your control as chairperson, but often they can be anticipated. Some aspects to be considered include the following:

• An ideal meeting room (which is rarely achieved) should have these characteristics: It should not be of the large auditorium type; it should not have a high stage; it should be equipped with a lectern, microphone, and large screen; it should have proper acoustical properties; and it should be arranged in such a way that speakers will be at a table in front of the audience, with discussants at a facing table if possible.

• You as session chairperson should meet with speakers and discussants fifteen minutes before the session begins, to review details of microphone, timing, instruction in slide projection from the podium (if controls are provided), lights, and any special requirements. At this preliminary meeting you should introduce all speakers and discussants to each other. If feasible, you might meet very informally with participants in your hotel room during the evening before the session (provide refreshments).

• The projectionist should be available in the meeting room before the session to receive slides and instructions (unless slides are to be deposited in a central office in advance of the session), to project any desired material under actual viewing conditions, and to *test the equipment to be used*. It is of great importance to have a trained projectionist, thoroughly familiar with the equipment to be used in the session. Unfortunately this rarely happens. Presentations of many otherwise good scientific papers have been ruined by faulty projection of visual material. (If the meeting is to be held in a hotel and you are driving rather than flying, you might plan to bring your own 2×2 projector with an extra bulb and extension cord.)

- Microphones should be tested in advance, and a flashlight pointer should be at hand and should be tested. A microphone that hangs around the neck, while a little awkward, is much more desirable than a fixed microphone if the speaker wishes to move around, perhaps handling slides, models, or other visual material. You as chairperson should determine in advance how to put on and take off the microphone and should instruct speakers in the correct procedures. If the room is large, one or more floor microphones should be provided.

- The chairperson should have a prepared introductory statement, to cover the period when the audience is settling down. Your remarks should include reasons for the session, ground rules, humor, and other general material. If at all feasible, the program should be constructed to provide you with this short period (up to ten minutes) for discretionary introductory statements, to be compressed if the session starts late for any reason (and it usually does).

- The chairperson should announce plans for coffee breaks at the beginning of the session—and their exact length should be announced just before they begin. If possible, a *long* coffee break is preferable to a short one. We all recognize that contacts made and discussions held informally during scientific meetings can be just as important as the papers presented. A long coffee break (possibly a half hour) makes possible informal discussions, which can usually be completed before the session is recalled to order. It is more convenient to have coffee served in the meeting room instead of in the corridor or in another room. Some clear signal should be given when the coffee break is over and the session is to be reconvened (even then, people will usually drift back in as they please).

• One of the most important but most consistently ignored functions of a session chairperson is the proper introduction of speakers. It would seem logical, and at least a matter of common courtesy, for the chairperson to provide the name, professional address, and complete topic title for each speaker—but even this bare minimum is often not achieved. The speaker's name is mumbled; his or her co-authors are not mentioned; his or her affiliation is often not given; and the title of the presentation is abbreviated or otherwise garbled. Faced with this reality, how can I go on to suggest that proper introductions are vital to the success of the session? Chairpersons *must* do the expected things mentioned above, but they *should* do much more. They should have some relevant biographical information about the speaker, they should know something about the speaker's research background, and they should have some comment on the research to be reported. Session chairpersons *might* even essay a small amount of humor.

• The timing of papers *must be absolute*. Speakers should be warned in advance about the total time available, then given a five-minute warning, then cut off when their time is up. Often, though, the pressure of time is eased by the last-minute failure of one of the speakers to appear. If that happens, you should announce the changed schedule at the start of the session and periodically thereafter, since some members of the audience may be there primarily to hear the missing paper.

• One of the real tests of the chairperson's effectiveness is the handling of discussion after the paper is presented. Audiences often sit mutely, and some spark must be provided to get discussion moving. When you are confronted with seeming unresponsiveness, one of the best devices is for you as chairperson to ask the first question. Often this breaks the silence. Other approaches are these:

1. You might assign a discussant for each paper. If so, the discussant must have read the full paper in advance of the meeting, and should have prepared comments.

2. You may select a panel of discussants for a discussion session which is convened immediately after the papers have been presented. Each panel member should be expected to either comment on or ask a question of every speaker. (For this elaborate procedure, a substantial segment of time must be allowed, and the presentation of papers must not infringe on the time allotted to the discussion.)

3. You might try "planting" prearranged questions in the audience—preferably with someone familiar with the work on which a particular paper is based. Such questions can be insurance against total silence, and should be used only where spontaneous questions or comments are not forthcoming.

4. One device to elicit participation from an audience is the solicitation of written questions. Small memo pads can be provided by the meeting hotel, and several volunteers designated to collect written signed questions, to be handed to the session chairperson at the beginning of the discussion period. The method is somewhat ponderous and formal, and does not promote the sometimes stimulating give and take of discussion from the floor, but it can be combined with a request for oral comments or questions.

- The chairperson should be sure that anyone asking a question or commenting on a paper identify himself or herself in advance. Questions and comments must be kept within the time schedule, and must not interfere with the presentation time for the following paper.

Occasionally, certain individuals from the audience may try to dominate the discussion, or worse, may launch into a long presentation of their own data—an impromptu paper. You as chairperson should maintain only minimum patience in such situations before halting the offender abruptly. More often than not, however, the reverse situation prevails, and it is extremely difficult to generate a lively discussion.

Sometimes, too, a member of the audience will persist in asking a series of obviously hostile questions, and may attempt to cite his or her own unpublished data leading to conclusions different from those reached by the speaker. A little of this give and take is tolerable, since it may provide some added life to the session, but it can quickly get out of hand and must be controlled by the chairperson. At some reasonable and early point the discussion should be shut off firmly but politely, by encouraging the combatants to meet privately in the alley after the session, or by the insertion of a little humor to defuse the situation if possible.

• *It is the responsibility of every chairperson to make succinct concluding remarks at the end of the session.* No session should ever be allowed to terminate abruptly or to dissipate haphazardly after discussion of the last paper. Your concluding remarks should briefly recapitulate the purpose of the session, the quintessence of the papers presented, and conclude with thanks to participants.

SESSION FOLLOW-UP

Once a session has been concluded and the participants have dispersed, your functions as chairperson are seemingly at an end. A few additional chores should, however, follow the meeting:

• You should assume responsibility for a letter of thanks to each participant—speaker or discussant—within a month after the meeting. If the session seemed successful, or if particular speakers made an outstanding contribution, specific comments should be included in the individual letters.
• If significant information, insights, or suggestions emerged from the session, they should be transmitted to the meeting chairperson or organizer, or to the society's executive board. Such notations are particularly valuable when discussions are not recorded or otherwise reported. Also, you might include suggestions about future sessions of a similar nature.

- You should assume responsibility for receiving papers resulting from the session if the papers are to be published in a collection. Usually the session chairperson is responsible for preliminary editing, before the papers are sent to the program chairperson or to the editor of the proceedings. Often one or more speakers may fail to submit papers on time, thus delaying the whole process of publication. You should take initial responsibility for "needling" tardy authors. For hardcase procrastinators you may have to resort to such embarrassing tactics as supposed phone calls from publishers or printers demanding the tardy manuscript. There is satisfaction, though, in seeing papers from the entire session in print—the logical end point and reward for the effort invested by the chairperson and the participants.

The chairperson of a scientific session can make a substantial contribution to its success by assumption of full responsibility. It is important, therefore, that guidelines and suggestions be provided to those who may be called upon to serve as a chairperson, especially if it is to be for the first time. In addition to the personal scientific attributes (acquaintance with many other individuals in the field, stature among colleagues, general familiarity with current research activities as well as the body of pertinent literature) that contribute to the effectiveness of a session chairperson, a number of other suggestions, some of them quite mechanical, may help in conducting a pleasant, relaxed, and fruitful session. Experience in the role is a valuable asset, but it can be augmented or replaced by a number of standard suggestions. Some of these, gleaned from observations at a great variety of scientific meetings, have been incorporated in this chapter. Preparatory actions of the chairperson can be equally important, if not more important, than activities during the session. An expenditure of time and effort in planning, preparing, and communicating will yield significant returns.

SPECIAL ASSIGNMENTS

As a small appendage to this chapter on chairing a scientific session, we should consider three special categories of assignments: The "keynote" speaker, the banquet or luncheon speaker, and the session (or meeting) summarizer. Each of them requires a high degree of skill, more than a touch of humor, and a smattering of game playing.

The *keynote speaker's* role is the most difficult to classify and define. We could get some insights by asking a series of critical questions, such as the following.

What is a keynote speech?

Nobody knows—least of all the keynote speaker.

Who should be picked as a keynote speaker?

Someone from far away with a lot of slides, and someone with "perspective"—which is another way of saying that he or she hasn't done any real laboratory or field research for a long time.

How long should a keynote speech be?

Long enough to make sure that all late risers have time to get to the meeting room before the substantive part of the session—the technical papers—begins, but not so long as to delay the coffee break.

What have you a right to expect from a keynote speaker?

Wit, charm, humor, pithy statements, broad and usually untenable generalizations, prophesies, a justifiable fear of confronting specifics, a desire to be loved and admired, and maybe, once in a while, that rarest of flowers—a fleeting glimpse of a fresh synthesis.

These are things you have a right to expect—of course, what you expect and what you get may be quite different things, depending on the ability of the keynote speaker. There are as well other questions for which there are no firm answers, such as: "Why are keynote speeches always so early in the morning?" "What relationship if any exists (or should exist) between the keynote speech and the rest of the scientific papers presented at the meeting?" and "What happens when the keynote is off key?" Your responses to these questions would be undoubtedly as good as mine.

Keynote speakers are usually paid (an honorarium, or at least travel and living expenses), and therefore more should be demanded of them than of the average meeting participant or author of a contributed paper. The letter of invitation to the keynoter should be very specific, outlining carefully what is expected and what is offered. If the keynoter is an administrator, he or she should be warned in advance to avoid making commitments, or statements which could be construed as commitments, but should be encouraged to bring any new tidbits of information from his or her agency.

The banquet or luncheon speaker fills a highly specialized role, and such roles are only for a highly qualified minority. The assignment is an impossible one anyway—to entertain, inform, and inspire a mixed group of people who have just finished a meal, which probably was preceded by a cocktail party. However, a speaker is part of the ritual, to the point where a group meal that just ends, with people drifting away after dessert, tends to leave a sense of unfulfillment or vacuum.

If by chance or design you are invited to be a luncheon or banquet speaker, consider the following:

- If *in your own estimation* you do not have a *significant* measure of *all* the primary qualifications for the role—wit, humor, perspective, and expertise—then *do not accept*. Do not be tempted or cajoled into a situation in which you and your audience will be uncomfortable or bored.
- Always make your talk *shorter* than expected, and *better* than expected.
- Keep your introduction and your windup *light*, but make sure that a core of substantive material is given. The entire speech cannot be fluff.

- Be very cautious about using visual aids as crutches. Many banquet rooms are set up abominably for projection, and part of your audience may not be able to see. If you do use visual aids, make them superb and make them few.
- Never read a speech, but don't wander into the room and ramble. Have a good outline, with an exceptional introduction and finale.
- Keep key points few, and splash them with relevant, crisp anecdotes.
- Don't even try to cover everything that you would like to say. Prepare a talk that seems reasonable for the time allotted, then take about one-third of it and do it well—discarding without a qualm the other two-thirds.

The banquet speech is an institution, and one which can have a significant place in scientific meetings provided it is treated with proper respect. Like most oral presentations it can be done superbly or abominably—or with great mediocrity. The basic operating tenet is that *people want to be and expect to be entertained at that point in the program.* They will tolerate and even accept a reasonable amount of substantive talk, provided it is framed in a light and pleasant context. The first few minutes of a banquet speech usually determine its fate, and form the basis for decisions by attendees as to whether they will stay with the speaker or begin preparations for a small nap. Thus these few minutes must be *very* carefully planned.

Humor must be handled with great care. Jokes are alright if they are clearly relevant to the topic and told well, with discretion and with good timing. Humor which is too obviously contrived may turn off the perceptive part of the audience. Humor which is too far off-color may be offensive to some, especially if the audience is mixed, and even if it is not. One good approach is to divide a talk into several fluffy and substantive subsections. The fluff can include jokes, anecdotes, and asides; the substance contains the message to be left with the audience—a message which should contain no more than three key points. Alternating between these two extremes, substance or fluff, after an attention-getting introductory few minutes, can provide a winning structure. Some speakers achieve this instinctively, but most must work hard to attain it.

The *session or meeting summarizer* (if such a slot is a part of the agenda) is probably one of the most demanding assignments at any meeting if it is done properly. To begin with, it requires the summarizer to actually sit through and even listen to every last scientific paper—a dreadful and often boring task. But this is just the outer chamber of hell; inside there are the real requirements of the job, which include:

- Ability to make astute and infallible judgments about the worth of each paper—an exercise guaranteed to make many life-long enemies;
- A lightning-fast ability to extract the real meat (if any) from the fat of papers, and to assemble the entire session (or meeting) package in logical order as quickly as shuffling cards; and
- A heavily keratinized skin, to view and accept with complete aplomb most of the audience's drifting away during your summary to catch planes; and to listen calmly to authors who were slighted or ignored in the summary, or worse still who felt that their work had been misinterpreted without an opportunity for rebuttal.

These major speaking roles—keynote speaker, banquet or luncheon speaker, and meeting summarizer—can add substantially to the pleasure of a scientific conference provided they are carried out excellently. To do so requires natural ability, careful preparation, and considerable experience. Such assignments cannot fall to amateurs or even semipros.

SUMMARY

Session chairpersons can contribute significantly to the success of a meeting if they become actively involved in the planning, execution, and follow-up. Selection and briefing of participants are important warm-up activities before the meeting. Adequate introductions and competent handling of discussions are crucial session responsibilities, and follow-up activities may include soliciting and editing session papers.

It is obvious that session chairpersons, keynote speakers, and other invited speakers constitute the hard core that determines the success or failure of many meetings. Good contributed papers are important, but they are necessary packing material in which the gems are embedded. Leadership roles in scientific meetings require competence, ability, and a high degree of proficiency in serious game playing. It is in this environment that all the early training and the instinctive responses of the professional player come to a focus.

CHAPTER 5

THE SCIENTIST AS A PRODUCER/DIRECTOR

Organizing Scientific Meetings

The annual meeting of a scientific society—the principal actors; innovations—please; the right kind of workshop; international science—participation therein.

After having served time in the trenches—presenting scientific papers at regional, national, and international meetings, serving on panel discussion teams, and chairing sessions at various meetings—the day will come when you will be asked to serve on the organizing committee for a meeting, or (worse still) will be asked to be the organizer or convenor of a meeting. Your natural positive response will be given quickly, without adequate consideration that the job will take a substantial part of your time for months, that the scientific benefits are vanishingly small, and that you will receive scant praise for your efforts, regardless of how successful the meeting is.

Nevertheless, organizing a scientific meeting is an eventual part of growing up in science; it has its own satisfactions, intangible though they may seem. Such a distinction is a marginal sign of making it in the scientific community, and a sign that at least some of your colleagues know you and trust you not to botch the job. The game player, if he or she accepts the role, will of course plunge in vigorously; do the very best job possible, even at the sacrifice of personal research; recruit able game players as a committee; and ensure adequate visibility of the efforts of all involved.

Since there are many categories of meetings, this discussion will try to consider only the principal ones: the annual meeting of a national scientific society, the workshop or small-group conference, and the international symposium. Obviously, there will be a number of factors common to all categories, but each type seems distinct enough to warrant special treatment.

THE ANNUAL MEETING OF A SCIENTIFIC SOCIETY

If you are "invited" to organize an annual meeting of a professional society, your cardinal rule is to work closely with and communicate constantly with the society's officers. The president considers this to be *his (or her)* meeting and you are just a useful functionary; so, while you are really in control, don't make it obvious that you are.

Beyond strict obedience to the cardinal rule, the next dictum to follow is to get started *early* on the scientific program. A year in advance is none too soon, particularly for selection of keynote and invited speakers, selection of session chairpersons, and decisions about format of the meeting (if there is flexibility in format, use it). Gather together immediately a small group of close colleagues whom you can count on to plunge in with you, make the selections above, communicate your plan to the president of the society, and with his or her approval make the first contacts with the proposed key people. Contacts must be made early, since there is a clear, direct correlation between percentages of acceptances and chronological distance from the actual meeting date. For people considered authorities, whom you wish for keynote, banquet, luncheon, or evening lecture slots, and whom you may not know personally, the best first contact can be made by one of their colleagues or former students. Initial contact can be by phone, but it must be followed immediately by a formal letter of invitation, which calls for a prompt written reply—for commitment or rejection.

Since we are talking about the critical backbone of the scientific component of the meeting—the major speakers and session chairpersons—there are a few game rules that might as well be inserted here:

- Regardless of the scientific achievements of any prospect for keynote, banquet, or other speaker roles, *if the person can't talk exceptionally well before large groups, do not consider him or her,* regardless of pressures that may be applied to you.
- You may be tempted to go outside the scientific field for a banquet speaker—maybe even so far as to invite a politician who has responsibility for or interest in some area of science. *Do not trust these people to appear!* They may have every good intention at the time of acceptance, but crises intervene, and a governor will send an unknown assistant, or a congressman will send an equally unknown legislative assistant. A good solution to this very real problem, since we would like key people to appear but can't depend on them, is to plan a dual banquet program—two speakers—with one speaker a scientist or science administrator (whose records of appearances are better) and the other a politician. Then when the politician backs out, often at the last minute, the banquet program won't collapse.
- A random survey of banquet speeches given at scientific meetings during the past five years disclosed a number

of fascinating findings: 1. Most of them are so ephemeral that memory of their title or content often does not persist even to the next day. 2. Many of them are too long or too technical for the kind of audience that will be present. Remarkably few speakers seem to give any thought to the nature of the mixed group sitting out in front of the head table, or to the reality that this is primarily a *social* occasion, and not the place for a scientific presentation. 3. A few of them are remarkably good, with a fine balance of humor and seriousness, with relevance to the purpose of the meeting, and with great consideration for the nature and physiological condition of the audience. 4. Some of the worst speeches are those given by some middle-level dignitary who has flown in that evening specifically for the occasion, who knows nothing of the audience or the meeting program, and who reads from a text prepared by his staff—often a text that he had not seen until he got on the plane.

In the end, though, selection of a banquet speaker is only a little better than throwing dice or picking random numbers. Someone who should do better may have an off-night; a relatively unknown second choice may perform like an opera star or a standup comedian. Success is important to the organizing committee. It wants every aspect of the program to be outstanding, but in view of the extremely feeble memories that most meeting participants have of past banquet talks, success or failure of the person behind the banquet podium will do little to change the course of science.

- Select your session chairpersons with great care, using some of the guidelines in the preceding chapter. Try to concentrate on people well known in the special field of the session, but also *people whom you know from personal experience to be capable and willing to conduct a smooth but dynamic session*. The job of session chairperson is not a passive, perfunctory one, and must not be treated as such in the meeting that you are organizing. If you must make a choice, select a person less well known but one whom you can count on to actually *lead* a session, over a well-known one who sees the role as simply one of introducing the speakers.

The format of many annual society meetings is often quite rigid, and unnecessarily so. This is particularly true of some of the older and more prestigious societies. Occasionally, though, at meetings of the board of directors of even the stodgiest societies there is some soul-searching about more effective and more dynamic meeting formats. Innovations are actually tried—some with success and some with failure. The basic dilemma seems to be that *while there can be no total elimination of presentations of formal contributed papers from the annual meeting, there must be a better way* (for at least part of the scientific program).

Among the innovative (in the sense of going beyond sessions of contributed papers) approaches that have been tried with some success are these:

- *Mini-symposia.* This device can be particularly effective in large national meetings with many concurrent sessions. At intervals throughout the meeting all concurrent paper sessions cease and a two-hour mini-symposium, with appropriate authorities, is held. After this, participants disperse to their concurrent sessions. This approach provides an element of cohesiveness that has otherwise slipped away from many large meetings.
- *Discussants.* One of the best meetings I attended recently made excellent use of discussants, in the true sense of the word. An entire day was devoted to invited review papers, and each one was followed by a shorter paper discussing the review from another perspective. (The discussant had received the review paper three months in advance.) Well-known scientists acted as speakers and as

discussants, and sufficient time was assured for less formal discussion from the floor.

- *Panels.* Panel discussions have been attempted with variable success for a long time. Many of them degenerate into successive presentations of formal mini-papers, with little time left for intrapanel discussion or interaction with the audience. Success of panels depends almost totally on the skill of the panel chairperson. Some approaches that I have seen succeed include the absence of prepared statements by panel members and the substitution of an introductory statement by the chairperson; the distribution to all panel members and the audience of a typed list of issues or questions to be addressed in sequence by the panel; and the use of two opposing panels, one of science administrators or politicians or environmental lawyers, and the other of scientists (differing world views can make the session remarkably stimulating).

- *Overview Papers.* At a recent national meeting every session was introduced by an overview paper and concluded by another one from a different point of view. The content of the contributed papers was known to the overviewers because of required early submission of abstracts, but the overview papers went well beyond the confines of those papers, and ensured that regardless of the quality of the contributed papers, there would be high points in the session.

Undoubtedly there are many more workable innovations that could lessen the boredom (speaking frankly) of hour after hour of fifteen-minute contributed papers. There must be a place for such papers, but they need not dominate the scientific component of a society meeting. The meeting organizer, especially if he or she is a game player and a gambler, would be well advised to innovate—and if the innovation doesn't work well, so what? A new organizer will appear next year to try something different anyway. *The purpose of the meeting is largely direct oral communication, and we are getting precious little of this in the standard meeting format,* so we have little to lose but our conservatism.

Beyond the early crystallization of the general meeting agenda and the selection of key actors, there are other important roles for the meeting organizer. One is close contact with the local committee responsible for arrangements (hotel selection, printing, reception, tours, projection equipment)—all the nitty-gritties which help to make the meeting move reasonably smoothly, and which unfortunately are used by many participants as criteria for judging the success of the meeting, and not the scientific content. Because this is a fact of life, assure yourself by personal observation that every last detail of local arrangements has been thought of and assigned to someone. This includes simple things like the nature of the badges and the size of printing thereon, whether Cokes will be available at the afternoon coffee break, what will be included in the spouses' program, how elaborate the hors d'oeuvres will be at the cocktail parties, transportation for invited speakers, the number of reception people at registration—on and on and never-ending details, each one important in making the meeting a perfect one.

There are other approaches to organizing meetings, beyond saddling a single scientist or a small group of scientists with the total responsibility. Some smaller societies have conned a few dedicated members into acting as a permanent program committee so that there is continuity from year to year at least in the scientific aspects of the meeting. Larger societies, fortunate enough to be able to afford a paid executive secretary, have invested much of the responsibility for the success of the meeting in that person, and in fact preparation and execution of the annual rites become a major function of the job. This too provides for continuity from year to year, and also introduces a stronger element of accountability if something goes wrong. Another approach is employment of commercial firms which specialize in organizing meetings and making all arrangements. They do not, of course, arrange the scientific component of the meeting, but do assume responsibility for just about everything else, including associated trade shows, which can absorb much of the costs of the meeting. Such firms operate on a contract basis. To become involved in this kind of operation, the society needs an alert and competent treasurer who can scrutinize proposed budgets and deal knowledgeably with financial reports of the meeting. Meeting registration fees go up of course, since these firms must be well paid, but it is a mechanism to remove the burden of details from the scientist.

Since these meeting-arranging companies will be a continuing fixture it would seem useful and important to focus on their positive and negative attributes for a short time. The people who head these companies are interesting mixtures of public relations types, often with a dash of the entrepreneur and the instincts of an accountant. They expect to be paid, regardless of the financial success or failure of the meeting itself—and ultimate financial responsibility rests with the society and not with the arrangers. Some rules of the road in dealing with such companies include the following:

- Every scientific society should be blessed, as I said earlier, with an astute, hard-headed treasurer who will follow with interest and understanding each step in the planning, conduct, and financial resolution of the meeting.
- The society should insist on a break-even budget from arranging companies—but should be prepared to see that budget destroyed by unforeseen events. Plan on at least a 20 percent overrun.
- The organizing committee of the society must never allow meeting arranging companies to get their hands on or to influence the scientific program in any way.
- The society treasurer should insist on, and should monitor carefully, an ironclad, exhaustively detailed contract with the arranging company.
- The greatest financial slippage can occur with tour and

banquet arrangements. No part of these arrangements should be left open-ended with the arrangers.

- A number of these arranging companies seem to be continually on financial thin ice. Don't enter a contract that gives them too much "up-front" money; they may disappear before the meeting date.
- Much of the work in arranging meetings is actually done by secretaries. Find out before a contract is signed just what the resident expertise in the company really is, and how much of it will be invested in *your* meeting.

Finally, and almost as a postscript, organizers of national meetings should be prepared to resist attempts to "politicize" large national meetings—either by subgroups of registrants or by outsiders. Politicization may take the form of circulation of resolutions or petitions clearly aimed at particular political groups or national blocs, and displays or exhibits that tout subjective views of scientific matters or national issues of the moment. Occasionally literature will be distributed, or corridor harangues staged, by people not directly involved in the meeting. Control of these extraneous events is difficult and should be left to proper authorities—hotel or police.

Probably the safest operating principle is to insist that scientific meetings confine themselves to science, unless it is clearly stated in advance, and understood by all participants, that broader objectives exist—such as formulation of recommendations or statements of concern.

THE WORKSHOP OR SMALL-GROUP CONFERENCE

Workshops with the right cast of characters are to many of us the best of all possible scientific meetings. I have participated in many workshop meetings that seemed extraordinarily successful. So, as a proper game player, I tried to analyze the organization and implementation of each one to identify, if possible, the ingredients of success: Workshops usually allow a much greater degree of interpersonal contact than do larger meetings, so this is a factor of success common to any workshop. But, beyond this, the ingredients that I see are:

- *The concept.* The conceptual plan for a workshop must be carefully considered, it must be clearly enunciated, and it must have sufficient breadth to fully engage the minds of all participants.
- *The mix of participants.* Depending on the conceptual base, the participants should be active, involved scientists, able to speak knowledgeably about their particular subdisciplines but also receptive to ideas from related subdisciplines.
- *The discussion leader.* An essential ingredient for success in workshops must be the discussion leader. Ideally, he or she will be an acknowledged leader in a subdiscipline of his or her field, someone able to lead but not dominate, someone capable of a light touch when the situation calls for it, and someone with good analytic abilities, able to winnow out a substantive and obligatory summary report from all the chaff of conversations.
- *The length of the workshop.* Unless the participants are well known to one another, a week seems to be an ideal length for a successful workshop. This allows each participant the expected time to establish his credentials in the eyes of the other participants, but then permits long, unhurried expanses of time for good exchange of information and ideas if the mix of participants is right. Such a time period also permits the development and review of a draft report—a very important product of any workshop.
- *The format of the workshop.* An outstanding recent workshop used the following structure: one day of plenary sessions, consisting of a statement of concept, objectives, and program by the convenor, followed by overview papers by each discussion leader and extensive discus-

sion in each of the subdisciplines contributing to the workshop objective; three days of subgroup meetings, with planned brief joint meetings of a number of the subgroups and the possibility of individual contributions to subgroups other than the one to which the participant is assigned; a final day of plenary sessions, in which the discussion leader of each subgroup was available at the front of the room for questions and comments (which proved to be extensive) on the report; and a final summation by the convenor.

- *The convenor.* The convenor's role in the planning of the workshop is of course very critical—especially in developing the underlying concept and in selecting participants. Beyond this, though, the convenor introduces the concept and objectives of the workshop on the first day and offers an overall summary on the last day. If the choice of discussion leaders and participants has been good, then they can be expected to perform effectively and the best contributions of the convenor are a continuing sense of purpose combined with a relaxed, congenial manner and prompt resolution of minor logistical problems that may develop.

- *The workshop location.* This item is low on the list of priorities since I have seen successful workshops in big-city hotel rooms and in delightful rural conference centers. The important common ingredient is that there is freedom from hassles about rooms, meals, travel, etc., and that adequate (even excessive) opportunities are provided in the evening for optional socializing in a single location. Once a group begins dispersing for sightseeing, shopping, etc., the productivity declines drastically.

- *The follow-through.* A frequently ignored component of a good workshop is the nature of its "follow-through." Pragmatically, any workshop is no better than its downstream results, yet this fact seems to be overlooked by a number of convenors, who seem content after their extravaganza to slip back into obscurity. Follow-through activities which should be considered include summary reports, public recommendations, specified public services, identification of leaders, proposals for future conferences, and publication of workshop proceedings.

The workshop, especially if it is scheduled for a full week, is a time of full exposure of every participant to a small group of his or her peers. Unless he or she is totally mute (and thereby a complete misfit in this situation), by the end of the week that small group will know in great detail the personality, the extent of knowledge of subject matter, the analytic and synthetic abilities, the conceptual depth, the sharpness of mind, and the attitudes of every member of the group. The workshop is not a time for games; it is a time when ability is clearly and starkly present or absent.

Taking for granted the existence of a high degree of competence, there are still some aspects of workshop activities which should be of interest to the game player. Workshop and symposium proceedings are often published as a book, and the *convenor* is usually the *editor* (hence has his or her name on the cover). Also, the overview papers, usually given by the discussion leaders, are often published under individual authorship as chapters in the book. So those jobs are clearly the ones of choice insofar as publication is concerned. They are not jobs that one usually applies for, but they are jobs that emerge from advance planning and preparation of proposals to appropriate scientific bodies. They all require work and preparation, and much personal contact, so should not be undertaken lightly.

The super-strategist in this editor game is of course the one who organizes a *series of workshops or symposia* and thus becomes editor of an *entire series of publications*—so his or her name appears on the cover of *every volume* and the editor of a specific volume retreats to the smaller type or vanishes into the "front matter" of the volume. In my entire career I have known only five individuals who achieved this absolute pinnacle of the "editor game" (and one of these owned the publishing company that printed the series). The poor author of an individual chapter or symposium paper to be included in such a series is almost helpless in the hands of these consummate strategists—until he or she learns how to organize a book-producing meeting, symposium, or workshop, and acquires the stature and contacts to carry it off well. The receptors of game players focus on the whole delicious process, and copious data are stored in internal computers for later integration into an individual game plan.

THE INTERNATIONAL SYMPOSIUM

Participating in the organization of an international symposium or comparable meeting—whether labeled "conference," "special meeting," "round table," or whatever—can be a pleasant and broadening experience, but it requires investment of substantial amounts of time. Usually, involvement in organizing an international meeting results from previous related activities at a national level, in society meetings, or in various kinds of workshops. Effective and enthusiastic participation at the national level, combined with a high degree of credibility as a scientist, often result in an invitation to help organize an international event, or in selection to serve on committees responsible for organizing the event.

To gain some perspective on the genesis of international meetings, it might be worthwhile to list a few examples:

- International federations of scientific societies in some discipline areas sponsor symposia at regular intervals (often every two to five years).
- A strong, discipline-oriented society in one country may make contacts with a counterpart society in another country and agree to co-sponsor a meeting.
- A number of United Nations agencies such as FAO, WHO, and UNDP sponsor or cooperate in sponsoring

regional or world conferences and workshops, on a recurring or one-time basis.

- A number of international commissions meet annually, and may organize special conferences or symposia to precede or follow the annual meeting.
- A government agency in one country, when confronted with a problem that is international in scope, may convene a special conference of world experts to review the problem and to make recommendations for action.
- A private foundation may underwrite a conference on a particular topic relevant to its goals or interests.

Through any one of these or other routes, scientists may become involved in planning an international meeting—often in association with bureaucrats from agencies supplying some or all of the funds. Although most meetings are unique, there are some common aspects which can be identified:

- The organizing committee will include representatives from several countries—some of whom will perform and others of whom will do *absolutely nothing*.
- If organizing committee members are selected from Eastern European countries, their participation will be minimal and communication with them difficult.
- In the rare event that a meeting is scheduled for an Eastern European country—usually because that country belongs to the sponsoring intergovernmental agency—it will be up to that agency, and not to individual foreign scientists, to interact with the internal bureaucracy of the country in planning the meeting. At best such meetings will be chaotic and frustrating. Meeting facilities and hotel accommodations will be "ornate-grungy," but rare opportunities for contact with good scientists from those countries may be worth the stress.
- Members of the organizing committee will often receive proposals for review papers or panel organization from

people unknown to them or their close colleagues. Discreet inquiry to colleagues in the country from which the proposal came is an approach of choice in this delicate area.

- The relationship of the chairperson/president of the congress/symposium to the organizing committee must be clearly defined at a very early stage. Of particular importance is the extent to which that person has veto power over committee actions—the degree to which that person's wishes become law. For most meetings the organizing committee may function as a board of directors, with reasonable but not total power.

- The financial commitment of each sponsoring organization must be agreed to in writing long before the first expense is incurred in the name of the conference.

Organizing a scientific meeting is really about money. The novice or unaware organizer who doesn't have this truism stamped on his or her forehead during the earliest planning stages of the meeting is a fool, or at least an idiot. For a long time during my apprenticeship in science I thought otherwise. My naive concept was that if the goals of the meeting were worthwhile and the prospective participants were credible, then the affair was certain to be, as if by magic, successful—sort of a "build it and they will come" delusion. Wrong. Wrong. Wrong. Any gathering of scientists that does not have a sound financial base from the very beginning is doomed to petty bickering, hurt feelings, and angry withdrawals by key participants. In accord with this thesis, the first meeting of any organizing committee must be about funding, and not about the scientific content at all. The good stuff comes much later, provided the dollar questions are resolved adequately (and not just tentatively or provisionally).

I have psychic scars and remote colleagues (who were formerly friends) to present as evidence for failures to accept my own printed advice about financing a conference. My most notable failure, just as an example, was an ocean science meeting in Puerto Rico. I was invited to organize a session on high seas fisheries research, with the understanding that territorial and private funding would be available and that all expenses of participants would be paid. I was still a journeyman gamesman at that time, so I didn't ask the critical questions about exactly who the sponsors were, and exactly how much money was available for expenses. I dutifully invited some very good scientists from southeastern universities to be participants in this proposed international meeting, with promises of simultaneous translation, funds for hotels, travel, and field trips, and a hard-cover book of proceedings. None of these things materialized. The funding was ephemeral, based on casual assurances from ocean service industries of "some level of support" that never became realities. Airline tickets were unpaid, the expected simultaneous translation service disappeared, most foreign participants cancelled at the last minute, and the meeting became a very local one-day affair conducted totally in Spanish, and concerned mostly with graduate student projects in Puerto Rico. Some of the people whom I had invited (and who had prepared papers that were never given or published) do not speak to me to this day. They blame me (correctly) for being gullible as well as naive, and these words, to scientists, are strong epithets as well as indicators of pitifully poor application of game rules.

- The chairperson of the meeting must submit a detailed budget for approval by the organizing committee, and every budget must project a financial "break-even" conclusion. Despite assurances of successes, organizers should anticipate deficits and near bankruptcy after any major international extravaganza.
- Be sure that a reasonably sound financial base exists for the meeting—from an international organization, foundation, or scientific society—before any public announcement is made and before you agree to participate. International meetings cannot dangle from shoestrings.
- The conference budget must be large enough to pay travel and living expenses of at least a few well-known

names as invited speakers. This can be critical to the success of the meeting since the presence of such names on a preliminary announcement will often be important in decisions that others will make about attending or not attending.

- Financial commitments in the name of the conference are the prerogative of the general chairperson, unless specific delegation to a member of the organizing committee or other financial officer is made.
- Plan an early and nonrefundable registration fee, to get an estimate of attendance and to be able to develop plans for expenditures.
- Set the registration fee high enough to cover a 20 percent overrun of the projected break-even budget and to allow for currency fluctuations.
- Be particularly careful of any early promises in preliminary brochures about simultaneous translation services. Prices are exhorbitant and are often fixed by union agreements. Unless carefully restricted, this could be one of the major items in the budget.
- Assume that at least one and possibly several of the invited speakers will back out, often at the last minute, even if their expenses were to be paid. Reasons may be trivial or substantial. Some speakers have a history of such behavior—easy acceptance and casual withdrawal—and should be avoided if the fact is known, regardless of their positive characteristics.
- Be *certain* that all invited speakers can speak *comprehensibly* in the official language of the meeting. If it is English, as is more and more the case, make sure that each invited speaker can speak "standard" English. Some slight accent is permissible and expected—but no more than slight. Otherwise the situation becomes impossible: the translators can't handle it and the English-speaking people can't understand. This criterion should outweigh prominence and scientific competence.

- Regardless of the scientific stature of the organizing committee members, a reasonable degree of nationalistic pride and protectiveness should be anticipated. Thus delicate matters, such as the meeting location, proportions of papers given by each country's scientists, equitability of funding commitments, official language(s) of the meeting, and national representation at the opening session and the banquet head table, all should be given careful consideration.
- Representatives from developing countries can be particularly sensitive about the status of scientific research and science laboratories in their nations. Cooperative programs with industrialized countries have been and are being planned and implemented, and much progress has been made in the past several decades, but sensitivities still exist and must be considered in planning meetings. Any patronizing attitudes should be strictly avoided.
- Once planning for the meeting begins, a major problem is maintenance of adequate communication among members of the organizing committee. Telephone and other forms of communication are excellent for some countries, but abominable or impossible for others. Even mail deliveries to some countries may take weeks, or mail does not get through at all. Telex and the Internet have proved to be methods of choice, provided all members have access to the equipment needed.
- As the meeting date approaches, the numbers of last-minute details to be dealt with multiply. Most or all of the organizing committee members have many other responsibilities, and may be away from their laboratories or universities for varying periods, often in locations where access to them for decisions or opinions is difficult. In these instances it is better if a "core" or executive committee, selected from the larger membership of the organizing committee, be designated to take actions for the entire group when necessary.

Despite frustrations and misunderstandings, interacting with scientists and bureaucrats from other nations in organizing an international conference can be a stimulating and usually pleasant experience. It is one of the best ways to understand some of the differences and commonalities in the practice of science in other countries, to develop some empathy for attitudes and philosophies of foreign scientific colleagues, and to develop good and lasting friendships with foreign counterparts. The game rules have added refinements, and the players are highly selected—all of which increase the pleasure.

SUMMARY

Organizing scientific meetings—whether they are annual meetings of a professional society, workshops, or international conferences—can be an interesting but time-consuming avocation. With the advent of paid executive secretaries in some societies, and with paid meeting-organizing companies proliferating, it is only the scientific program itself which must remain the purview of volunteer scientists. Since the scientific program is the core of the meeting, it must be well planned and superbly executed. Program organizers should take advantage of whatever latitude exists for innovation, and should suppress, to the extent possible, the usual but boring lineup of short contributed papers. Organizers should maximize opportunities for interpersonal contacts, with mixers, long coffee breaks, and hospitality suites underwritten by the society—any device to bring participants together informally.

CHAPTER 6

THE SCIENTIST AS A NEGOTIATOR

Participating in Committee Meetings

The committee chairperson's creed; role-playing in committee meetings; the value of committees; the routes to committee chairpersonship; advice on advisory committees; the inevitable document.

Committees are formed for many reasons. The only thing that they have in common is the existence of a chairperson, on whose success the success of the committee depends. Committees are usually advisory, although some may have decision-making authority (or think they do). Normally, an advisory committee produces some type of report, which may serve as the basis for decisions by others.

Look at the whole sterile paragraph above. Not a word is said about the complex interpersonal activities, the devices available to the chairperson, or the highly variable collective moods of the group. These are the ingredients of which committees are made, and to understand their functioning is a fascinating study, and properly an area of great interest to the game player.

THE CHAIRPERSON'S CREED

Since the chairperson has been identified already as the key to committee success, it would be sensible to begin there and try to identify some of the elements of success. Many chairpersons carry with them a small card on which, in tiny, tiny print, are the basic commandments of the job. Enlarged, the card looks like this:

The Committee Chairperson's Creed

I will guide, but I will not give the impression of dominating.
I will have a carefully conceived position, but I will be very careful not to reveal it too early in the deliberations.
I know the conclusion that I want to reach; I only need let others discuss the matter long enough with proper direction until they convince themselves that they have reached the same conclusion by group agreement.
On every key issue I will have an initial position and at least two fallback positions.
I vow never, never to let the group dissolve without having something on paper with which all members can agree.
I will insist that the committee function under rules of common courtesy and concern for all participants as people—that no one will ever be allowed to feel small or uncomfortable or excluded.

Some of the elements of this creed may seem at first glance to be overly manipulative, and even to deny the necessity for existence of the other committee members. Nothing could be further from the truth. The chairperson must have done his or her homework, and must have thought through problems adequately enough to have prepared positions, but must also be fully prepared to entertain an alternative position if it is good enough. The effective chairperson must be aggressive, but must also be perceptive enough to know when to shut up and let others move with the tide of discussion. The chairperson, most of all, must be able to bring the discussion to some logical conclusion and to record on paper a statement that the entire committee can live with, even if it is in rough-draft form. This must be done before the group evaporates; no right-thinking chairperson will allow dispersal of members with the vague promise that someone or some few will put together a document which represents the sense of the meeting, because it will rarely if ever happen.

The effective chairperson also recognizes that the tensions of small-group meetings must be relieved periodically by humor, coffee breaks, or other diversions. The perceptive chairperson should ensure that this relief occurs, and must pick the proper moment so that productive sequences (the "highs" of most meetings) are not interrupted—usually never to be fully regained.

Beyond these obvious elements which contribute to a chairperson's success, there are other good operating rules:

- Don't push members too hard for too long—remember that a committee meeting is a social occasion too.
- Expect brilliant performances from your key actors, and give them verbal cues and openings so they will know when to take the stage.
- Be prepared to fill any transitory vacuum in the flow of discussion with an astute summary of progress to that point, or with a relevant series of questions.
- Provide social outlets if the committee activity extends over several consecutive days. Informal cocktail parties, group dining, relevant field trips—all provide tension release and help mold a compatible entity.
- Don't try to be "on stage" every minute. Occasionally let the conversation proceed without your contribution or intervention—then after a suitable period, draw in the reins.

THE COMMITTEE MEMBERS' GUIDE

What of the committee members themselves? They are usually present because of some acknowledged expertise or point of view or position of consequence. How do they play their roles? *Experienced participants in committee activities will have evolved, consciously or unconsciously, definite and identifiable techniques.* Some will deliberately overstate their position or express some dogmatic stand—and then be prepared to back away. Some will not indicate their true position until all other members (or at least some key members or the chairperson) have expressed themselves. Some will always state a negative or pessimistic view (the nay-sayers). Some will always want to charge ahead full speed with whatever the chairperson proposes, or with *anything* which seems new or innovative. Some will spend all their time trying to impress others with their brilliance and wittiness, relevant or not. Some will consistently play the role of devil's advocate, regardless of their own convictions. Some will indulge in personalities, especially if the group meets regularly. Some will play the role of peacemaker and compromiser. On and on it goes! How is a poor chairperson ever to cope with all this role-playing and still get the business of the committee accomplished? (Answer: With great difficulty and monumental forbearance.)

The astounding thing is that under the right chairperson a mix of at least several of these types is salubrious, stimulating, and often productive of decent conclusions and actions. The value of committees should not be underestimated. They may be time-consuming, and their true worth may be partially obscured by seeming inability to reach a consensus, but the reasoned judgment of a small group is more often (but not always) a better judgment than might be made by any single individual. Also, committees provide occasions for valuable contacts with peers and department heads; they provide visibility, favorable or otherwise, for some junior members; and they provide an opportunity for all to gain insights about the value systems and mental processes of other participants—insights which may be important to the observant game player in situations outside the conference room.

Committee members have every right to know the guidelines and strictures under which they are to operate. If the committee is purely advisory, this should be absolutely clear from the outset. If the committee is empowered to make decisions, this should also be clear, as should the method of implementation of decisions and the authority on which implementation depends.

On the other hand, committee members should subscribe to certain operating guidelines as well. The most basic is that committee membership should be accepted only if active participation in the affairs of the committee is feasible and planned. To avoid a disservice to the rest of the committee, if you can't participate regularly, say so and decline the opportunity.

Another general operational rule is that acceptance of a committee assignment implies a *supportive* role in committee activities. Too often certain committee members seem to participate only in a destructive, belittling way—to deliberately reduce the effectiveness of the group and to eliminate the pleasures of interpersonal exchange.

Committee meetings can be critical testing grounds where supervisors form opinions about winners and losers, based on such evidence as articulateness, perception, analytic ability, knowledge of the field, and ability to relate positively to others.

The element of significance to you as a game player in all this is to enter committee activities vigorously and enthusiastically, but at the same time to stand back and enjoy the complex interpersonal game that is being played around you. This can make even the most routine committee meetings pleasurable and stimulating events. Observe particularly the role of the effective chairperson, because this is the best part of committee assignments if carried out with finesse, good humor, and intelligence.

The routes to committee chairpersonships are varied in the extreme, but can be paved by proper role-playing as a committee member. Elements of this role include avoidance of obvious attempts to dominate discussions; sensible, low-key, reasoned comments and suggestions; perception as to when to back off; and acceptance with some grace the defeat of personal proposals.

There is one other aspect of committee assignments that might be mentioned. Committees, if properly chaired, can sometimes be effective deterrents to dictatorial authoritarian tendencies on the part of department heads or others in decision-making positions—provided they are not allowed by their members to degenerate into rubber stamps for the wishes of those in control. This can be an extremely delicate matter, however, since few committees have any substantial mandate for continued existence and may be abolished almost instantaneously.

SOME COMMITTEE GAMES

There are almost endless devices and ploys that surface in committee or other small-group meetings; some are employed by the chairperson and others by the members.

One device used by the domineering chairperson is to organize the committee meeting to simulate a classroom situation, with the chairperson as teacher. The most common game in this situation revolves around the principle that "I know the answer and I'm going to draw it out of you." This game can go on interminably, with the chairperson in a position to give a pat on the head to some ("you're close"), and to belittle others ("you haven't understood the problem"). The great danger here is that adults will quickly perceive the game and resent being treated as juveniles. However, if the chairperson's power base is strong enough, the members will play along and will not often overtly challenge the chairperson, regardless of their inner feelings.

Another device that can be used effectively by the chairperson is the judicious selection of a rapporteur whose opinions generally parallel his or her own. Since the summary minutes of the meeting are often the only tangible residue, discussion that goes counter to the preconceived positions of the chairperson can be minimized or not even recorded. Sometimes these deliberate oversights can be corrected if all the participants have an opportunity to review the minutes, but often this is not done because of pressures of time.

Still another ploy used by some chairpersons is the personal selection of an *ad hoc* executive group or "kitchen cabinet" of close associates—an informal group that meets in advance of the committee meeting and selects positions on the issues to be brought up at the meeting. This device, if tolerated, diminishes the functions and contributions of the rest of the committee.

A final chairperson's ploy is to develop, by diligent effort during committee meetings, a complete spectrum of opinions on a subject, from strongly pro to strongly con. In the deliberate absence of consensus, the chairperson may, after the meeting, blithely follow the course that he or she had planned from the outset. Perceptive committee members watch for this device and press for some kind of consensus before allowing a topic to be dropped.

The committee members also come to the meetings equipped with their own devices. One of the most annoying and most difficult to counter is the practice of deliberately interrupting the chain of thought and discussion with an extraneous comment, or offering gratuitous negative value judgment about a suggestion being presented by another member before the statement is completed. A good chairperson will not tolerate too many episodes of this kind, but an imperceptive one may not act until other members are thoroughly intimidated or frustrated.

Another very interesting ploy that is used in almost every meeting can be called "establishment of turf." This can be done verbally as well as physically. Verbal position is achieved by carefully worded references to one's own expertise and accomplishments so there is no doubt in the minds of other participants. The physical game involves progressive lateral expansion of conference table space, with papers, notebooks, briefcases, laptop computers even projectors—so that the neighbors' table space is gradually compressed or eliminated altogether. Good players combine this with usurpation of space *under* the table with aggressive foot movements and insertion of briefcases. A higher order of this game calls for the deliberate change of position at the table after coffee or lunch, to take another's seat after he or she has established a temporary claim. These devices annoy and upset other participants, thereby reducing their effectiveness relative to that of the game player. It is always surprising how much of this petty and often deliberate abuse some participants will take before exploding; in the interim their attention to affairs of the meeting is badly distracted.

Still another device that appears in committee meetings is the stated need for a member to leave early for some presumed high-priority reason. If the excuse is accepted by the chairperson, this permits the transient to present his viewpoints and opinions on agenda items early and without interruption, and does not allow sufficient time for proper rebuttal or discussion. If the committee later tries to reach a consensus different from the opinion of the now-departed member, it is faced with a dilemma which may involve postponing any action.

It seems that each committee meeting brings out a new variation of the games played by its chairperson and members, and this makes a fascinating and never-ending study for the astute game player. Recognition of the game being played can be a source of satisfaction, but too much attention to games can, if not kept within reasonable boundaries, quickly reduce effective contributions to the mission of the committee.

COMMITTEE WOMEN

Although the topic of women in science will be explored in a later chapter, some mention of the effects of women scientists on committee meetings might be made here. Female members of committees present a whole new set of challenges to men, who are used to interacting with other men in such settings. Some of the challenges are the necessity for men to observe and learn a new female body language (about which books have been written); a felt need to study any new types of games that female members of the group may be playing (as distinct from the familiar male committee games described earlier); and a vague unease about revised role-playing engendered by the presence of female colleagues. Naturally, until territories and operating guidelines are established there may be minor friction points—just as there are in all-male groups. Female members will adopt roles, and will be categorized, sometimes unflatteringly, by male members. Favorite categories to be expected are the "pushy broads," who are self-assured, knowledgeable, and capable of moving in on any discussion; the "chip-on-the-shoulder females," who seemed poised to pounce on any remark that can be construed as sexist; and the "token female," who may have moved rapidly through the system and who may not yet be comfortable with the new environment.

Whatever the price, the frequency of participation of knowledgeable females in committee meetings is increasing. I recently attended several committee meetings in which men were in the minority, and even some that were chaired by women. Such trends can be disquieting in some conservative scientific circles, which are still largely male-dominated. Regardless of the composition of the group, though, committee meetings can be productive, pleasant, even stimulating social events. Committee members should approach them enthusiastically, anticipating challenges, successes, failures, and put-downs—but more pleasure than pain.

THE ADVISORY COMMITTEE

There is a very special category of committees in science known as "advisory committees" or "advisory boards" that deserves particular attention by game players. They are the committees which have the greatest potential for making impacts, but they can also be just rubber stamps for agencies or organizations. Members of an advisory committee are often selected because of some kind of eminence, power, or visibility. A matter of great concern is that such a committee may disappear quickly, before its initial advice is translated into action, if it is too much at odds with established agency or organizational positions. Another concern is that advisory committees can be allowed to wither and disappear slowly by deliberate agency inaction, even if the committee functions adequately.

The game player can find great satisfaction as a participant in the activities of advisory committees. Properly manipulated *by the committee members*, advisory functions can be productive and meaningful in shaping policies and agency positions. Advisory committees need not be facades, unless they allow themselves to be manipulated by the creating agency or organization. Perceptive and aggressive game playing by even a minority of the committee is all that is required to prevent abuses.

A colleague recently described her participation in the charade of an advisory committee created totally as a "front" for a federal environmental agency. The committee was created ostensibly to assist in planning and evaluating pollution-related research projects being conducted largely by contracts. The reality was that these planning and oversight functions were performed entirely by the in-house staff, and the committee met so infrequently that its impact approached zero. The executive secretary of the committee was a member of the agency staff, and minutes of the rare meetings were not distributed to the committee members. Then it was revealed that the only reason for the committee's existence was to satisfy some vague dictum from higher administrative levels.

All these facts emerged slowly. Fortunately a few members of the advisory committee were game players, and they proceeded to reverse

the tide gradually, to avoid the risk of sudden elimination of the com-
mittee by the agency. Some of the approaches used were insistence on
agendas in advance of the meeting, insistence on approval of sum-
mary minutes and lists of recommendations before each meeting
adjourned, public release of committee advice and recommendations,
increase in frequency of meetings, requests for participation by news
media representatives, and the conduct of workshops which led to
public statements and memoranda. The agency was clearly uneasy,
but was reluctant to kill the committee in view of its public visibility.
In the end the committee's advice to the agency proved useful and
important.

PAPER PRODUCTS OF COMMITTEE ACTIVITIES

Committees and committee meetings are ephemeral, and
all participants know this. Therefore there is often a subliminal
urge to leave behind some legacy, some record of a brief hour
on the stage. The legacy usually takes the form of a document—
a report—bristling with conclusions and recommendations,
and vaguely representing some of the topics discussed during
the meeting. The creative chairperson and/or rapporteur can
have a moment of glory in producing such a report, especially
if notes taken during the discussions are selective and not too
detailed.

European scientist/administrators have long known what
is only now beginning to penetrate the consciousness of United
States counterparts—that the meeting report is important, in
fact that the meeting exists only to produce a report. There
must be a document produced from every meeting; without it
there has been no meeting, only an informal discussion. The
policy is a good one to remember and imitate, since the report
is the only tangible residue of many hours of work, and since
something of significance may have been said. (Of course some
reports remind one of the mouse produced by the pregnant ele-

phant after long labor, but that is a hazard which must be accepted.)

Because of the potential significance of the committee report in ways which may not be perceived immediately, it seems important to offer a few guidelines:

- Next to the chairperson's job, the role of rapporteur is the plum. Good committee reports extend beyond merely recording or summarizing discussions and recommendations; they result from skilled action by the rapporteur—an act of creation without subversion of the sense of the discussions. A good rapporteur is the product of years of observation and practice.
- The value and importance of a "creative rapporteur" should never be underestimated, since the notes and summaries made by that person are often the only permanent residues of a meeting. Products need to be selective, concise, and unambiguous—and need to be available almost instantaneously. The rapporteur's job is a power position, second only to the chairperson's, but to utilize the position fully a good rapporteur must be a *participant* in the discussions and not just a *recorder* of events. One way to free time for participation is to develop a shorthand method of writing down only key statements or decisions, and then to dictate soon after the meeting a summary report built on the shorthand notes. Another approach is to transcribe shorthand notes into a draft summary during the late-evening hours—and in the case of long meetings, to do this consistently every night. To do these things well requires practice and some natural talent. The satisfactions that can be derived include the ability to slant or weight arguments pro or con on an issue, the opportunity to cast particular individuals in a favorable mode and to ignore or minimize the contributions of others, and the opportunity to work closely with and to

study the thought processes and maneuvers of good chairpersons.

- Good committee reports can result from on-the-scene participation in their preparation by members of the group. Plenary discussions may be alternated during the day with the production of draft written statements by working subgroups, so that by the end of the meeting the chairperson and the rapporteur have the rough outlines of a document and accompanying verbiage. With minimal polishing, a draft final report can emerge from these working documents. What is even more important is that most members of the committee feel some personal involvement in the report.

- If the report is at all extensive (beyond ten pages), it should be preceded by a brief "executive summary" for those without the inclination or time to read the entire report. Such a summary constitutes an extremely difficult but important piece of writing: it must contain the major points made in the report; it should contain a minimum of jargon or bureaucratese; and it must sell the program or proposal.

COMPUTER CONFERENCES

For those who secretly or openly resent the time demands of committee and staff meetings, there is a recent development in electronic communication that brings relief—the computer conference. With its roots in the old rural party phone line and in conference phone calls, the computer conference has added the niceties of printed retrievable messages, a memory for comments inserted at 3:00 a.m. or any other time, instant recall of the words used and the sequences of arguments and rebuttals—all done remotely, from the relative privacy of office or laboratory.

This new type of "absentee conference" comes with good

news and bad news. It is an enormous time saver if participating members are geographically separated—even across town. Inputs can be made sporadically, interspersed with periods of attention to other activities or inactivities. Members are not constantly "on stage" as they are during committee meetings or conferences; they can even make contributions from their bedroom. On the negative side, the absence of face-to-face exchanges obviously lessens human impacts, and the chairperson's job is made more complex by drastic reduction in control.

Uses of computerized conferences are almost limitless. They can be used in refinement of the successive stages of a planning document by geographically dispersed staff members. They can be used to explore options and to reach group decisions.

Scientific conferences, workshops, and symposia, along with conferences generally, are gradually adopting computerized approaches that try to retain some of the flavor of personal interactions but avoid some of the time and travel restraints. I had a chance recently to participate in a test of "multiple site interactive video conferencing" as part of a marine pollution meeting held in Charleston, SC, and sponsored by the federal National Oceanic and Atmospheric Administration (NOAA). The other sites were a regional pollution assessment meeting in Seattle and a meeting of NOAA pollution program administrative staff in Washington, DC. Technical papers were presented at each site and participants from each site joined in the discussion. No major glitches developed during the entire teleconference, and discussions with remote site participants were judged equivalent to what might have occurred if all participants had been assembled at one site. A number of interesting comments were overheard during the virtual coffee breaks between sessions:

- *Interactive video conferencing is an excellent vehicle to bring regional issues (such as, in this example, pollution in Puget Sound) before a national audience of experts, and to provide detailed regional presentations not otherwise easily accommodated at national meetings.*

- *Despite the obvious advantages—and to be weighed against them in any evaluation—interactive video conferences of this type at the present state of the art are still equipment and labor intensive ventures, involving literally dozens of support people at all levels. As the technology improves, the extent of hands-on technical support will likely, of course, diminish.*

- *Maybe it is from attending countless meetings and conferences in one specific location, but video conferencing does not seem to provide the "community of scholars" atmosphere and the free "give and take" that are to most participants essential ingredients of scientific meetings.*

- *Interactive video has obviously great prospects for future workshops, especially those that are short, but intense. When combined with e-mail, the drafting of joint proposals, progress reports, final reports, manuscripts—almost any kind of multiauthored document—can be markedly simplified.*

So, whether we are ready or not, a piece of the computerized future has invaded the halls of science, insofar as the classic gatherings of societies and the mass migrations to symposia and workshops are concerned. I feel that we need to adapt, and to test new approaches, but we don't need to give up all our favorite playthings to the great god progress.

From the point of view of the game player, the presence of a strategically placed computer terminal as an integral component of office equipment provides evidence that the occupant is in intimate contact with the universe. The pro will, however, have a secretary or technician who is responsible for most communications using the new toys (but he or she will have full competence to handle the equipment personally).

The ramifications of electronic conferences are awesome. As a start, air travel for scientific purposes may be reduced drastically; international symposia may be replaced by a series of subgroups seated behind individual keyboards and facing a giant video display panel; and scientific papers, reviews, and even books may be written by co-authors who have never met in person, but who may have worked together remotely for years. The debit side includes the risks of greater isolation of individual scientists, loss of stimulation from personal contacts, and deterioration of oral communication skills.

Speaking personally, I, who readily admit to anachronistic tendencies, hope that some of these new approaches are only fads. Science is done by people, and opportunities for scientists to interact should not be suppressed or unduly channelized, since such interactions are among the most important career satisfactions for many of us.

SUMMARY

Committees and committee meetings have been variously described by scientists as curses, necessary evils, stupid time-wasters, occasionally productive window dressing for bureaucrats, rubber stamps, and training fields for novices. In truth they are all these things and more—they are important test sites and practice courts for interpersonal skills. Committee chairpersons, rapporteurs, and members all have roles to fill—roles which often seem to follow prepared scripts. Good chairpersons are perceptive, diplomatic leaders; good rapporteurs must be active participants as well as selective summarizers; and effective committee members should be supportive, involved, and articulate. All must be aware of and subscribe to game rules if the group endeavor is to be productive. A principal criterion of committee success is the written document which emerges— a document which captures the quintessence of the discussions but rarely the nature and depth of human interactions.

PART TWO

CRITICAL ISSUES FOR SCIENTIFIC STRATEGISTS

Basic training is over. The fundamental interpersonal strategies critical to success in science have been sketched in broad outline in Part One—all the way from writing the first scientific paper to taking the lead role in advisory committees. It is time now to focus on some less clearly defined but absolutely crucial aspects of people interactions that can mean victory or defeat. After much agony, I have reduced these critical components to three—transitions (moving up, on, and out), power, and ethics. These to me are the core elements in the thinking of any scientific strategist. The manner in which these three components are approached and integrated shapes much of the productive life of any practitioner of science.

Here then, in Part Two, is the heartland of the scientific strategist. Here is where actions are taken and decisions are made that influence—in fact often determine—the course of an entire career. Choices about jobs—where and what kind—should be those of the scientist and not of the system. The scientist can and should be more actively involved in the decision-making, as a player rather than as a pawn. Once he or she has achieved professionally, then the sources of power become available; the uses of that power constitute learned strategies for the few. The ethical base of scientific practice is a matter of concern for all who consider themselves "scientists," yet too few examine this base in terms of their own activities and attitudes. The strategist will have analyzed the ethical foundation of his or her discipline and will have reached conclusions about where in the ethical spectrum his or her operating principles reside.

Part Two, then, is a consideration of decisions made by scientists—where they will work, what jobs they will consider, what uses they will make of opportunities to control others, and what ethical principles they will use as guides. These decisions will affect the entire professional life of any scientist, so they need to be made with great care.

CHAPTER 7

THE SCIENTIST IN TRANSITION
Moving Up, On, and Out

Winning strategies—how to be promoted; the fast track—what it is, how to know when you are on it, how to know when you are being pushed off, when and how to get off voluntarily; the life stages of a scientist; eras in a professional career; losing strategies; moving on—that special feeling of wanting to go and wanting to stay; factors critical to decisions about moving; moving out rather than on; retirement.

Success in science is usually measured in part by quality and quantity of production, much as it is in many occupations. Probably the most fundamental difference between scientific and other kinds of work is that the arbiters of success in science are more frequently *external* to the organization for which you work. These include peers, "authorities," editors, manuscript referees, society officers, granting-agency officials, grant-proposal reviewers, and site-visit team members. The "outside" judges are superimposed on *internal* authorities such as immediate supervisors, department chairpersons, laboratory directors, deans, faculty review committees, and promotion review boards.

Fortunately there are enough objective criteria for "good science" to make the decision-making process about your worth as a scientist a relatively simple one. Commonly used criteria include sustained productivity, knowledge of the literature in the research area and adjacent to it, ability to reach reasonable conclusions from available data, ability to make warranted syntheses, ability to perform reasonably sophisticated statistical analyses of data, and a record of publication of reviews or books.

Unfortunately there are also many subjective criteria which impinge upon definitions of "good science." These can cloud the basis for decisions about your worth. Some of these are location and nature of graduate training, your relationship to the "in-group" in your specialty, kinds of journals in which you publish, participation in society affairs and scientific meetings in your specialty area, the extent and nature of your informal correspondence with peers, your personal relationships with colleagues, and the relative success of your former graduate students or assistants.

It seems clear that winning strategies in science must take into account both objective and subjective measures of worth. Judgments about you will not be made exclusively on the basis of objective criteria, regardless of attempts to apply formulas or weighted evaluation systems. The realities of this situation are accepted by game players.

It should be inserted here, though, that a significant part of the reward of doing good science is in the internal satisfaction and the sense of accomplishment that comes with exploration and exposition of hitherto undescribed phenomena, or with the development and proposal of new methodologies or new syntheses. These are the creative acts that make a scientific career so satisfying; these are also the things that are less vulnerable to subjective judgments by others. It is always important to add these internalized achievements, evaluated by your own criteria for accomplishment, as leavening to the judgment of peers.

Beyond these internal phenomena, however, it is entirely reasonable to look for external signs of success. One obvious and visible sign is, of course, *promotion*. Most people like to be promoted—for a variety of reasons, such as greater prestige, higher salary, or more authority (or combinations of these). Promotions in science do not come from fundamentally different sources than in other occupations; the descriptive words are different, that's all. Some simple guidelines are:

- Assemble, quickly, a substantial publication record in a research area which, by fortuitous or planned selection, has high current visibility, rapid payoff, and great fundamental or practical significance.
- Be willing to move frequently, then to adapt quickly to the new situation and to begin producing immediately.
- Develop and express openly a genuine sense of involvement with and loyalty to the current job situation and your current supervisor or department chairperson.
- Remember that people with openly expressed enthusiasm are easier to see at promotion time than the excessively reserved.
- Keep in mind that a modest amount of aggressive behavior, indicating to appropriate supervisory/administrative levels that you are eligible for and are anticipating a promotion, is not unexpected, if it is done in a frank, open, nonconfrontational manner.
- Assiduously and absolutely avoid any formal public disagreement or confrontation with superiors, even if you are right; and tiptoe gently, diplomatically, but firmly into any private conversations to explore areas of disagreement.
- Remember that "political" skills are the real determinants of success, and not merely scientific excellence. Living and advancing by political means is seen by some as a threat to personal value systems—if the term is considered to embrace too many negative connotations (manipulation, lack of compassion, cunning, deceit, cor-

ruption). The reality is that personal integrity and ethics need not be challenged or violated by sensible acts that can be described as "political."

Some research situations, particularly in those specialized areas which are undergoing rapid expansion, can be highly competitive, and call for a reasonable degree of aggressiveness. There is still room, however, for kindness, tact, and courtesy. These characteristics will be appreciated and remembered long after the research has faded into back issues of journals. Among those who remember may be some of those evaluating you for promotion.

THE FAST TRACK

There are, of course, many routes upward beyond that of producing significant research. Scientists with ambition, energy, and ability can rarely be completely satisfied with themselves until they have tried the "fast track." Of course, the same can be said of able individuals in other professions—law, clergy, engineering, etc. Examples of the fast track insofar as scientists are concerned include (but are not limited to) such jobs as a position with a granting agency or large government laboratory; a position with significantly broader, often nonscientific responsibilities, such as legal, managerial, financial, or public affairs; a position with an international agency such as the Food and Agriculture Organization of the United Nations (FAO), the World Health Organization (WHO), or the World Bank; a position with a large private foundation involved in scientific, cultural, or humanitarian work; or an academic deanship.

Often scientists find themselves on the fast track without consciously electing to enter the race. Decisions that lead individuals in that direction are often made early in their scientific careers—and in some cases unknowingly. The shifting of gears from science to administration has been so smooth as to be almost unnoticed, or has been sensed but ignored.

In recent decades, coincident with the increasing complexity of scientific projects and programs, and with the proliferation of funding sources, there has been a proportionate increase in movement of scientists from research and teaching to management and administration. The products of this movement constitute a spectrum of job categories lumped under the title "science administration." The spectrum ranges from a relatively junior scientist, recruited early to serve as a minor functionary in a granting agency, to a senior executive with a science background but with many present responsibilities beyond science in areas such as finance, public relations, and even political interactions.

The "bench" scientist often views such people with dismay and disbelief. To him or her it is beyond comprehension that someone trained in science would deliberately abandon research and teaching to become involved in seemingly sterile pursuits that are usually two steps or more removed from the "real" world of science, which is in laboratory or field research. Yet discussions with science administrators at all levels make it clear that there are challenges and satisfactions to be found in management and administration that can for some be equally as great as those in research. Trying to describe these challenges and satisfactions is a difficult task, and often ends up being remarkably unconvincing.

- Some scientists are clearly more people oriented than others, and they tend toward managerial positions in laboratories and organizations where their principal role is in personal interactions rather than research.
- Some scientists have exceptional analytic abilities, and they find great satisfaction in grant reviews and program evaluations, activities which lead them toward administrative jobs with granting agencies of the federal government, or private foundations.
- Some scientists are more issue and policy oriented, and tend toward administrative jobs with activist organizations or private foundations.

- Some scientists are politically aware and interested in the political process as it affects science. Such people often gravitate toward state departmental positions in areas such as energy or environmental protection, while others move into legislative assistant or consultant positions.
- Some scientists simply want to make money, so they form consulting firms which, if successful, require movement away from the practice of science.

So for these and literally hundreds of comparable reasons people with scientific training sooner or later move from science to science administration (and occasionally back again).

LEAVING THE FAST TRACK

Once you are on the fast track it is very important to know when you are in danger of being pushed off into the surrounding underbrush. Some of the warning signs and the pitfalls to be avoided include the following:

- Your rate of progress through positions of increasing responsibility, complexity, authority, and salary must be constant, with no major plateaus in excess of four years.
- You must never move laterally to a position of similar responsibility and authority, and under no circumstances, either as a result of internal reorganization or merger, can you afford to occupy a position of lesser authority and responsibility—this then decides the future course of your career, and must be avoided at all costs.
- As you move into and through science-related but essentially nonscientific positions, the matter of personal relationships with others inside and outside the organization increases in importance exponentially. No longer can you depend on scientific productivity and the scientific judgment of peers as criteria for your advancement—

rather you must now depend on such intangibles as favorable personal contacts, impressions of your worth gained by others after only fleeting association, political interactions, and how you have treated others in previous positions you have held.

It is fairly easy to adjust to such a quicksand arena, but it does take conscious effort to avoid deliberately antagonizing those who might now or later be called upon to pass judgment on you, or to contribute to a decision affecting your future. The best approach to people and problems seems to be one characterized as straightforward but not necessarily blunt; open, but not naïve; authoritative, but not dictatorial; and knowledgeable, but not obnoxiously so. An excellent maxim is to *treat every person you contact with respect, since any one of them may later reappear as your supervisor or judge.*

Beyond this maxim, at higher administrative levels success is a mix of being in the proper place at the proper time, saying appropriate things at appropriate times, and knowing and interacting with the proper people. Simple, isn't it?

Sooner or later many runners elect to get off the fast track *voluntarily.* Some do so because they find themselves falling back from lead positions while others fail to find expected satisfactions. Among groups such as administrators with scientific backgrounds the act of voluntarily leaving the track is not an easy one. Often loss of prestige and drastic reduction in salary are involved. Often, too, scientists-turned-administrators find their scientific skills obsolete if too many years have intervened since publication of their last scientific papers. They must suddenly reoccupy a place in a scientific world peopled in part by younger, highly selected recent Ph.D.'s with fresh new information and techniques. Fortunately the feelings of inadequacy should be short-lived, since an effective scientist who is familiar with literature sources and who has maintained some contacts with his or her peers can quickly bridge the years given to administration or other nonscientific occupations.

LIFE STAGES OF A SCIENTIST

By examining the life histories of many scientists, the conclusion emerges that there is an extremely logical evolutionary process at work, forming a continuum with at least six identifiable stages: (1) the novice, (2) the journeyman, (3) the professional team leader, (4) the science manager, (5) the science administrator, and (6) the senior scientist. Figure 1 illustrates the stages in the continuum, plotted against scientific competence, and the following brief sections are summarizations of personal diary entries from one scientist who traveled the whole route.

Stage 1. The Novice Scientist

A fresh new Ph.D. from a good university, a first job with some degree of freedom, colleagues tolerant of mistakes and enthusiasm—and the new scientist is on the trail to glory. Sure there are minor problems, like heavy teaching loads or low

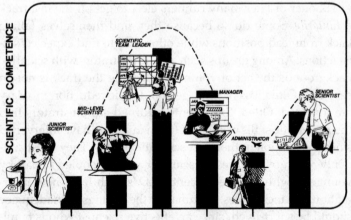

FIGURE 1. Stages in the life history of a scientist, plotted against relative scientific competence.

salary, but to function at last as a professional is one of life's genuine highs.

Stage 2. The Journeyman Scientist

Fifteen papers and one promotion later, the mid-level scientist is still sure of the correctness of his decision for a career in science. He has joined societies, presented papers at meetings, and even won a prize for best paper at an annual meeting. His graduate student load is heavy, but sharp, energetic students help maintain the early personal enthusiasm.

Stage 3. The Professional Team Leader

The research area selected so carefully for a thesis has been expanded far beyond earlier expectations. Research grants, mostly federal, have allowed purchase of elaborate equipment, and his space needs have been at least partially met by the institution. In addition to what seem like hordes of graduate students, he now has a secretary, grant-supported technicians, and several exceptional postdoctoral fellows. His peers seem to like him too—since he has been elected to head a national society and his grants get renewed with some regularity.

Stage 4. The Science Manager

The decision to assume the directorship of a private research laboratory was a wrenching one, since, despite initial hopes, it has meant almost total abandonment of direct involvement in research. The satisfactions come now in planning and evaluation, and all the personal interactions with scientists that are necessary for the maintenance of a productive research organization. A new and demanding role has emerged.

Stage 5. The Science Administrator

Increasingly, laboratory management functions involved contacts with representatives of funding agencies, including the federal government, so when the opportunity arose for a modestly senior agency administrative job, it was taken, despite the housing costs in Washington. Decisions about large-scale funding of university research are a significant part of the responsibilities, as are policy and planning functions. All these have proved challenging, as have frequent interactions with legislators and industry representatives.

Stage 6. The Senior Scientist

Feelings of accomplishment in administrative activities were coupled with those of increasing remoteness from the day-to-day affairs of science, despite valiant attempts to keep up with some scientific literature and even attend an occasional society meeting. Eventually the day arrived for a decision to accept a senior faculty position with total freedom to teach or not teach, and to do research and writing on topics of his choice. The return to scholarly life, completing a circle, was a satisfying final stage in a full and rewarding career.

Some move along this continuum of life stages "scarce knowing if they wish to go or stay," while others—the game players—make deliberate choices and deliberate moves to propel themselves in the direction they elect to take.

There are only a few principles here, and even these are a little shaky:

- Once the path is chosen, it is difficult (but not impossible) to reverse direction. It is more feasible to reverse direction in a university setting than in a government or industry laboratory.
- Once the path is chosen, a series of physical moves should be expected, since the upward path tends to be slower in a single laboratory or department, with a limited number of niches at higher levels.
- There are "way stations" along this curve where a scientist may pause temporarily or stop off permanently—at his or her own volition.

- Because of limitations in competence, insight, or energy, any point on the curve may become an involuntary plateau.

Examining this continuum, it seems to me that the process of conversion from scientist to manager to administrator is a complex one deserving of careful study. One clear principle to follow is to avoid playing at any level as an amateur. As soon as you find yourself supervising even a small group (or sooner), take some short courses in supervisory skills and personnel management. As soon as you find yourself managing a group (involving a reasonable degree of decision-making and financial control), take some university management training. A number of universities now have excellent seminars, short courses, and in-service training institutes, some of which are especially designed for science managers. The final step in the metamorphosis—to science administrator or science executive—can also be enhanced by some wide-ranging advanced training, and again a number of universities and other institutions have developed policy-level courses of all kinds specifically for scientists with an urge for retreading. Selection must of course be based on the position responsibilities and interests of the retread, but can range through such titles as Science and Public Affairs, Budgeting, Science and International Affairs, Public Relations, Organization Structure, etc.

If there is one failing that I have identified in the conversion process from scientist to administrator, it is the unwillingness of the convert to admit that management training is a critical part of the transition, and (even if it is admitted) unwillingness to carry through with the training. Of course a high degree of native ability, intuition, perception, and concern for people can carry a convert far, but when so much can be *learned* about good management, it is ridiculous not to take advantage of the fund of available information.

With the proper preparation for what is essentially a new career, although one built in part on accomplishments in a previous career, the scientist-administrator often finds pleasure, challenges, and satisfactions in his or her new role. "Management" and "administration" are not necessarily sterile jobs designed for people with short imaginations. These roles can combine continued interest and participation in science (though not necessarily at the bench level) with broader experiences in planning large-scale and long-term projects, with manipulations to ensure funding of projects, with maneuvers to ensure the best staffing of projects, with implementation and evaluation of the research, with inspiration to produce excellent results, with public interactions to ensure that the results are known—in short, with all the broader components of science. These can be exciting and fulfilling roles for science administrators, and many converts have found them to be exactly that.

Probably the most important iteration that I can make here is the one implicit in every chapter of this book—*whatever the game is, play it as a professional.* Nowhere is this dictum more relevant than in the transition from the familiar, comfortable laboratories of science to the executive suites, where an entirely new set of game rules apply. These new rules are beyond the parameters outlined for this book, but fortunately there are other rule books for managers and executives to turn to when you are ready. Even though countless books have been written about management as a science and even though many management techniques can and should be *learned,* successful management of *scientific* organizations is still to some extent an *art,* representing a dynamic mix of professional credibility, managerial skills, perceptions, instincts, and energy.

At this point it may be useful to the strategist to look at some samples of scientific managerial types that have been successful, in the sense that scientific productivity of the group was above average—which must be the final criterion. Successful science management types range from the "scientific dictator," authoritarian to the core, to the "delegator-chairman of the board," an extreme example of participatory management—with intermediate types which can be identified as the "no-

nonsense production boss," the "grand old man of science–father figure," and the "thinking person's manager."

The point is that there is no single successful management type or style in scientific organizations. Scientists expect reasonable management procedures, and as a group they can adapt (sometimes by selective attrition and replacement) to any form of management. Scientists usually want to be consulted about research projects in which they will invest their time, but other aspects of participatory management seem to be of less concern to them.

ERAS IN A PROFESSIONAL CAREER

While the pseudo-Shakespearian concept of the six ages of a scientist just presented has reasonable validity and is only moderately associated with the chronological age of the individual, there are other useful ways of looking at a scientific career that are tied more closely to time measured in post-Ph.D. years. One that has some appeal to me divides a scientific life span into three eras or decades (assuming a statistically average career of three decades): the crucial first decade of professional life, the uncertain but critical middle years (decade two), and the mature scientist (decade three).

1. The *first decade* may include some or all of the following:

- The all-important first job and the major intervention of luck and timing;
- The important first grant or contract;
- The first crop of graduate students;
- Paper presentations and other participation in scientific society activities; and
- Successful and pleasant relationships with colleagues.

This decade is a fast-moving one, filled with a sense of urgency and often with good feelings of professional accomplishment. A new family makes demands, and the struggle to maintain some sort of balance between a first love (science) and personal commitments outside the laboratory and classroom is difficult. There seems to be just too little time to do all things well, and it is often the family relationships which suffer. Scientists are often mediocre or poor spouses, because of addiction to their jobs—but then, this is true of many occupations which demand total commitment.

The first decade is also a time of high mobility, as better positions and salaries become available in other institutions for those who emerge as winners. It is also a time of greatly expanding scientific horizons, as expertise broadens beyond the solid but narrow base developed in graduate school.

With emphasis beginning in the early 1980s on federal budget reductions, and with the market for traditional positions persistently weak, we are seeing a proportional increase in post-doctoral research positions and junior nontenurable faculty appointments. These temporary slots may occupy part of the first decade of professional life—but they offer additional challenges for good junior scientists. Also, with the advent of complex but less stable private-sector ventures (such as genetic engineering companies) the range of opportunities for the well-prepared, aggressive entry-level scientist seems to have expanded.

2. *The middle years* are a mix of most of the following:

- Maintenance of scientific productivity despite increasing burdens of science-related activities;
- Successful grant strategies;
- Right decisions about moving;
- Impact of family stresses on science;
- Midlife crises for male scientists;
- Science versus administration—turning points;
- Sabbaticals; and
- Books and reviews.

This second decade of a scientific career builds of course on successes or failures in the first, but is often characterized by a continuing series of critical decisions—such as exploration of a new field of research, how best to spend a sabbatical year, whether or not to accept a "fast-track" job outside research and teaching, and whether to invest time in writing a book.

This is often a time of frank personal assessment of competence and credibility in a chosen field, and a time of assessment of these factors by others in decision-making jobs. Promotions, elections to society offices, and invited lectureships are clues to the outcomes of such assessments.

It is a time, too, of maturation of professional relationships. Every subdiscipline in science has an internal system of communication and support among those considered to be producers and leaders; to be part of the system is important to any scientific career.

3. *The mature scientist* may find the following milestones in the third decade:

- Importance of internal satisfactions;
- The big external rewards—prizes, department chairmanships, leadership of scientific societies, editorships;
- Maintaining competence and productivity—avoiding scientific obsolescence;
- Friendships among scientists;
- Planning for retirement; and
- Disappearance in a self-determined but brilliant exit scene.

The third and culminating decade of a scientific career has wide variability, but can be the most satisfying and exciting of all. Here the peaks in all aspects of science are available—editorships, invited participation in international symposia, visiting professorships, academic administrative jobs, participation in grant reviews, and leadership in advisory groups.

Often the extent of "hands-on" research is reduced in favor of directing research teams, or planning and evaluating the research of others.

Depending on individual preferences, the later years of the decade may be spent in a "senior scientist" category, with relative emancipation from schedules or requirements imposed by others, and freedom to summarize and integrate the experiences of an entire career.

There are internal satisfactions and dissatisfactions in every era. The first decade can be a happy one, but often passes in a blur of achieving; the second may be characterized by the accomplishment of some career goals, the frantic pace of administration, and the disarray of a mid-life crisis; the third may be a blend of maturation of intellect, success or failure in larger stadiums (national or international awards, dean-ships, society leadership), and good or poor health.

It seems important to note that while there is some statistical validity to the concept of a thirty-year postdoctoral career, many scientists are productive well beyond this average—particularly in recent years, with increases in retirement ages and some diminution in mandatory retirement policies. The third "decade," therefore, needs to have a very flexible outer boundary. Whatever its actual length in years for the individual scientist, it often represents a period when career accomplishments peak and then blend gradually into a final stage of summarization and integration.

THE "PROFESSIONAL"

I have wandered through countless scientific meetings, sitting through endless technical papers, consuming innumerable cups of coffee at breaks, and drifting from group to group during conference mixers (cocktail parties)—always with the same question: "What is a professional?" My conclusions, based on all this research, are subjective of course, but I do see some common identifying features that set the true professional apart from the average utility-grade practitioner of science:

- The professional has done substantial research and writing in his specialty—enough so that he has "credibility" among colleagues. The professional also has "visibility" because of his excellent oral presentations, the content and nature of his published work, his ideas, and his ability to relate positively to peers and others. The professional stimulates others and is often an innovator. His conversation stretches the mind without fracturing credibility. However, he rarely dwells on his own work in casual conversation except in the abstract, or in response to direct questions.
- The professional is "accessible," and is never too busy or pressed by a schedule that he cannot spend a few minutes in the hall with a graduate student or colleague. He is, however, a frequent traveler involved in a matrix of society meetings, advisory committee meetings, seminars, invited talks, review board sessions, course lectures, and a host of other time- and energy-consuming activities in which he is a significant if not dominant force.
- The professional is usually overcommitted to too many projects with too little time, but if he makes a commitment it will be carried out fully and often superbly.
- The professional transmits a certain "joy" about the science that he or she is pursuing—a quiet or outspoken pleasure to be just where he or she is in time, place, and vocation.
- The professional persists. Stress tolerance and energy levels are high; organizational upheavals are confronted calmly but not passively; and career setbacks are accepted with grace, recognizing their temporary nature.

- The professional usually has a substantial amount of that ephemeral quality called "charisma," a complex mixture of favorable personal characteristics, absence of pronounced negative mannerisms, innate or learned skills in interpersonal relations, multiple enthusiasms, an active mind, quick wit, and instinctive ability to respond correctly in time of crisis.
- The professional knows that he or she is good, and insists on continuing personal excellence.
- The professional is extraordinarily perceptive about human interactions and the consequences of every action.

Scientists with this mix of characteristics are not abundant, but they do exist and can often be found as society officers, department chairpersons, laboratory heads, or research division chiefs. The novice game player has much to learn by studying identified specimens of this superior animal in as many habitats as possible.

As a postscript, during my aimless wandering in search of "the professional" I stumbled upon three other categories of scientists (subsets of the professional, but with subtle recognizable differences). These I have labeled "the survivor," "the chairperson," and "the gentle manipulator."

"The survivor" is a member of a special and very elite class of individuals who seem to emerge undamaged, smiling, and smarter from the worst that any system can offer. My observations of scientists who are also survivors disclose these common characteristics:

- Survivors are people of controlled enthusiasm: they are involved, but they maintain a small measure of emotional distance in any interaction.
- Survivors are not prostitutes, but they know instinctively when to emerge and when to withdraw—when to fight and when to lie back and enjoy.

- Survivors have nearly infallible instincts; they are part of the action but always somehow escape real injury.
- Survivors are superbly perceptive; many of their actions are learned responses to perceived situations.
- Survivors are not surprised at being used—in fact they expect attempts at manipulation, and are prepared to respond with skill.
- Survivors take defeats with grace, and always arise from disasters, phoenix-like, with feathers intact.

Look for survivors in the dying moments of conference cocktail parties (they're the ones still sharp and alert, even after five martinis); look for them as major contributors of summarizing information on the last day of a week-long workshop; look for them to emerge with a decisive statement that shapes the final opinion of a review committee; look for them as leaders of very late-evening discussions in hotel rooms during scientific meetings—in brief, look for them somewhere beyond the point where others have long since given up and disappeared. They are, clearly and unquestionably, my favorite people.

Next in order of favor (with me) is *"the chairperson,"* a scientist who, regardless of the situation, quickly rises to a position of control and power. He or she

- is bright and aggressive, but not to the point of turning the average person off;
- makes good contributions to group actions, whether invited or not;
- volunteers for difficult and time-consuming assignments, and carries through;
- understands "the system" thoroughly, and avoids proposals or actions which disturb that system;
- automatically assumes the center of activity and discussion at a small-group meeting; and
- has unusual ability to form a cohesive unit of people with disparate views, and to lead them to the creation of a

good product (planning document, meeting summary, research proposal, or set of recommendations).

The third and final subset of "the professional" is a category which I have labeled *"the gentle manipulator,"* although this term may have excessively negative connotations. The gentle manipulator

- has a reasonable and genuine degree of scientific competence, but doesn't depend on it for success;
- is often a careful "user" of people, and during waking hours (and probably even when asleep) is planning future manipulations of people and events;
- is smooth, urbane, charismatic, with a perpetual but not blatant self-interest;
- can talk his or her way past guards, police, or attendants—all of those with instructions to exclude the average person;
- is able to get choice window seats at a restaurant while others stand in the street;
- is given, and expects to be given without hassle, an oceanfront room at coastal meeting hotels at no increase in price while others get a view of the parking lot;
- quickly finds a cab on an otherwise deserted and rainy street in Seattle;
- is able to get the ultimate in free and eager services from otherwise harassed airline cabin attendants—far beyond the normal call of duty;
- finds a water taxi at 2:00 A.M. on a small canal in Venice; and provokes a smile from a restaurant waitress in Moscow.

The recognition and proper description of the professional, and subspecies thereof, is to me a significant achievement of the sometimes insignificant research on which this book is based. Association with people who are true professionals is one of the privileges and pleasures of a scientific career.

LOSING STRATEGIES

Thus far this chapter has been upbeat, in accord with part of its title, "Moving Up"—emphasizing winning strategies in movement through the life stages of scientists. There is, of course, a darker side, characterized mostly by dead-end situations, that needs to be touched upon very lightly. To represent the losing side, I have selected three candidates: the wrong choice of a first professional job, the gopher syndrome, and persistent mediocrity in science. I have other candidates—such as the obligate windmill-tilter and the hollow strategist—but for the present I'll go with the following three:

The Wrong Choice of a First Professional Job

It is a truism that there is probably no more critical decision in the professional life of a scientist than that concerned with the first job after graduate school. Often the entire course of a career is set by that decision. Whether the employer of choice is university, government, or industry, the decision is surrounded by danger. Of course, with the present job market for many categories of scientists the options may be much narrower than the ideal, but this is the way I see it:

- Earn the Ph.D. before leaving graduate school even if you are on the edge of starvation.
- Make an all-out effort for a job at a well-known college or university; second-rate colleges will overwhelm you with teaching loads and usually discourage research, despite catalog material to the contrary.
- It is difficult to escape from junior college or high school teaching once you have compromised your ambitions to

that extent; a marginal postdoctoral somewhere is a much better choice—at least you'll be alive scientifically even if you're still starving.

- Entry-level staff jobs in government agencies for new Ph.D.'s are often—no, usually—stultifying, resulting in quick erosion of fragile and hard-won expertise because of the absence of any hands-on practice of science.

- Temporary positions with government research laboratories are reasonable gambles, since you are then on the inside, and there is always the possibility that a permanent job will open up for the qualified and aggressive person. (There is also the danger that it won't, of course.)

- Entry-level industrial research jobs are often as technicians rather than as professionals (regardless of the title). Stay in such a slot only as long as you are learning, and expect a high turnover rate among newly hired colleagues.

- Jobs with consultant firms are high risk and often short term, for the duration of a contract. Recognize them for what they are.

The Gopher Syndrome

One of the most clear-cut losing situations in science is to find yourself drawn, by whatever chain of circumstances, into what I (and others) call the "gopher" or "Rumpelstiltskin" syndrome. The sickness exists in most laboratories and university departments—in fact in all types of scientific organizations. One or more individuals, usually of junior rank, are tagged as "gophers," to do much of the "scut work." They are usually pleasant, cooperative types, who soon find that their personal and research time has been eroded to the vanishing point by the gradual acquisition of chores that logically should be done by others, or at least spread around. These chores include such items as responsibility for the departmental seminar series, briefing new teaching fellows, maintenance of projection slide catalogs and even equipment—on and on, item by item. In small research groups one graduate student often emerges as the gopher. He or she may win or lose in that situation depending on the tact with which the job is handled and how much time is left for science. It constitutes an early, severe, and somewhat unfair test for an incipient game player.

Persistent Mediocrity in Science

An idealistic view of scientists is that they are as a group totally dedicated, fully committed, without concern for diversions outside their profession. Not so! Too many individuals identified as scientists are content with mediocrity and search diligently for ways to avoid real involvement in scientific work. Hobbies become all-important, even infringing seriously on that part of the day that should be spent in productive research. Endless hours are spent in plotting strategies to avoid professional responsibilities. Enthusiasm, if any, is directed away from the practice of science.

The basic error here is in referring to such individuals as "scientists." They are *time-servers*, and they populate any field, professional or otherwise. They may have drifted into science as a relatively unstructured, poorly supervised area of work where their minimal efforts are often tolerated or overlooked. Science is just a "front" for these people; they expect the advantages and the income, but they contribute little or nothing. Their published scientific contributions are minuscule or trivial, usually totally at odds with their own self-perceptions.

It is disconcerting that such vocational misfits may adopt some of the manners of game players and thus usurp the form without the substance of science. They may learn and attempt to practice some of the game rules while refusing to pay the basic entry fee of sustained scientific productivity. They may even succeed for a time in obscuring their hollow core with a facade of pseudo-professionalism.

These deliberate nonproducers do a disservice to science and to scientific game playing—as long as they are able to maintain the masquerade. They want the profits without doing the work, and for this they must be exposed. Good, productive scientists who are also expert game players are best equipped to identify and destroy the phonies and pretenders—and should do so with great vigor and relish.

MOVING ON

A special subset of the topic "Moving Up" is logically that of "Moving On," at any point in a scientific career. Scientists are a reasonably mobile group as professional groups go. Skills are often transferable, although really good job openings have diminished and may remain scarce except possibly in the fields of environment, biotechnology, energy, population control, and medicine. Moves to other positions constitute major decision points in any career and are not to be taken lightly. There are some legitimate game rules that are relevant to such decisions.

- When the opportunity exists for a move which seems beneficial and advantageous from a career viewpoint, be sure that you assess personal and family factors fully. Such factors must be given as much weight as career factors, and in fact they may be overriding.
- The question sometimes arises, "Can I get ahead faster by making several moves?" The answer is clearly and unequivocally yes and no. "Yes" if you are a mediocre or average, or even slightly above average, scientist. If you are mediocre, moves should be made before too many of your peers or supervisors become convinced of your mediocrity. Once this happens you may roost in a static situation, in terms of salary and rank, for a long, long time. If you are average in productivity and effectiveness and the right opportunities present themselves, it may be

possible to increase your salary and/or rank by judicious changes in location. If, on the other hand, you are good or even outstanding, it is less necessary to move frequently. A productive scientist should be able to move through the system (provided there are enough niches at all levels) in one research organization or department and in one location if he or she so chooses.

- An exception here is the new Ph.D. who is asked to remain at the parent institution as an instructor or assistant professor. Such a person, regardless of his worth or accomplishments, will always be thought of and treated as a glorified graduate student, and he or she will watch others—often less qualified—move in and then move beyond them. The lesson is a simple one—*never, never, regardless of inducements, stay more than one or two years at the institution which gave you your advanced degree.* After you have made your way in the outside world, however, it is perfectly proper to accept a call to return to that campus in some reasonably exalted position.

- Often a move is made or contemplated because of inability to coexist with a supervisor or department chairperson. This can be a serious problem, since that person is in a position to influence your rate of progress (if any) in the organization. Diplomacy, tact, and private discussions on areas of conflict should certainly be attempted, but if conflicts cannot be resolved in this way, or if you cannot feel a continuing honest sense of loyalty to the position or the people, then you must leave. There is absolutely no other acceptable course.

- Try your utmost to leave a scientific position with the best possible feelings, and devote some effort to maintaining contacts, since the members of the group you are leaving behind will be important to you for a long time. They will be consulted, or they can offer opinions on your potential, your demonstrated abilities (or disabilities), and your role in group situations. These people can have an input on decisions affecting your career long after they have ceased to have firsthand information about you.

- Do not join the ranks of the wanderers, the dispossessed, too precipitously or for inadequate financial reasons alone. The good, productive scientist has more options than administrators would have him or her believe—such a person *can* stay in one place and grow scientifically and personally. The old saw that you must move to progress is a deliberate falsehood perpetrated by those whose success depends to a large extent on convincing people to do things they don't want to do at a time when they really don't want to do them.

- The decision about leveling with your present supervisor or chairperson about job-seeking plans is a delicate one. He or she will probably be consulted anyway by your proposed new boss, so great secrecy may not be that important. On the one hand your present supervisor, especially if he or she is inexperienced, may be shocked or resentful of the fact that you want to move. Life from the moment of disclosure may become slightly uncomfortable for you. Experienced supervisors, on the other hand, expect a certain amount of turnover, and take such matters calmly. Furthermore, if you are considered worth keeping, your treatment and general job situation may actually improve. I think that if there is a choice, it is probably more sensible to reveal than to conceal.

- In job-related discussions, don't be afraid to be a little assertive—even aggressive—but not obnoxiously so, and not to the point where your supervisor or manager is embarrassed. Job-related discussions should always be treated as business matters to be conducted in a business environment, and not at social gatherings (especially not at cocktail parties). A private luncheon appointment and discussion is alright if this is the way that particular manager conducts business (and some don't like to use their lunch hours for such purposes).

- Of course, there must be a tiny element of judiciousness about moving. If a laboratory or research unit is being reduced in size or discontinued, there is little professional advantage in being a holdout after the season starts. There is an advantage, however, in not overreacting to rumors or suggestions about what might happen in the future. Often the situation is so unstable that a whole pattern of events which seems assured at one moment may change radically, almost overnight, as a result of new funding and new programs—so hang in there for a while.

- There continues to be fluidity in movements between jobs which are primarily teaching and those which are primarily research. The net movement favors research, because salaries are better and job-related stresses can be lower. The dominant stream flows seem to be younger high school, junior college, and college science teachers into technician and entry-grade professional jobs in industrial research or into management training, and mid-level, often older, government and industrial scientists and science administrators into university, college, and junior college teaching, often on a part-time basis at first.

An interesting case of cross-matching was described in a recent Sunday supplement article on mid-life career changes. A scientist from a large industrial research complex had elected to become a full-time junior college teacher, at less than half his previous salary, after twenty years in research and research administration. A search for new challenges and satisfactions (a typical mid-life crisis response) was the reason given for the change.

Almost simultaneously, a young assistant professor of physics at a nearby small private liberal arts college was making a reverse transition to that same industrial research complex, to an entry-level posi-

tion with a salary still substantially more than he was making after seven years on the faculty. Slow advancement, lack of clear-cut tenure decisions, and insufficient research facilities at the college were the stated reasons for the change.

Both these individuals were obviously good, productive scientists, but both were unhappy enough with their existing professional niches to undergo the trauma of a radical job change.

With the present very dynamic climate of increased private-sector responsibility and reduced federal support of research—combined with explosive development of totally new high-technology industrial activity—decisions about proper positioning to take advantage of future opportunities are difficult to make. More than ever before such positioning for younger scientists should start during graduate studies. Deliberate moves to short-term postdoctoral positions with strong research groups may be approaches of choice—producing long-term career advantages. Temporary job instability and poverty can thus be traded for more rapid entry into complex private-sector research.

MOVING OUT

As a postscript to the matter of "Moving On," and possibly a subject as important in its own right, is the matter of "Moving Out." Such drastic moves—leaving science altogether—are not uncommon, especially today when alternate lifestyles are being explored as never before—lifestyles which often lead to satisfying alternative careers. Some scientists leave the field to make money, since academic scientists as a class are poorly paid in this country. Some scientists leave because the expected joys and satisfactions of scientific employment have simply not been realized. This seems like a tragedy and is difficult for most of us in science to comprehend, but it is a very real problem for a few.

Regardless of motivation, however, moves out of science, if they are to be made, should be made before age forty, and preferably to an alternate job for which some preparation has been made. Some scientists have produced and marketed their own products, based at times on ideas developed during their scientific employment; many others have expanded hobbies into full-time businesses; still others have tried scientific farming; and some have entered industry at managerial levels.

A category of "Moving Out" that most scientists face eventually is of course *retirement*—gradual or rapid withdrawal from science as an occupation, in later years. There are many things that could and should be said about the process, but there are so many variations that it is difficult to create convenient categories for descriptive purposes, as I have done with other phases of a scientific career. There are a few generalizations, however:

- Good scientists rarely retire abruptly and completely. They may request reduced teaching loads in favor of more extensive writing; they may move from research and administration into consulting; or they may expand scientific avocations—but they usually don't disappear completely from the corridors and meeting rooms of science. This is probably because most of them genuinely enjoy the careers they have selected, so retirement is not necessarily an important goal.
- Since the practice of science is mostly mental rather than physical, chronological age is of small importance. Good science depends in part on accretion of background and information in subject matter areas. This process goes on throughout an entire career, and often reaches fruition in the later stages of a scientific career.
- In many cultures, present and past, the elders play and have played a very important role in transmission of history, ideals, and ethics to newer generations. This is still an important function for older scientists in our society—

not to be truncated by too-early disappearance. Most older scientists take this responsibility seriously.

- Immediate preretirement years are often periods of summation and integration, producing books and reviews, or elaboration of concepts, that can only result from prolonged and intimate involvement in particular disciplines.

- A recent phenomenon in industrial and government science is earlier than usual formal retirement, with the simultaneous assumption of academic careers, often part time, and often in graduate schools. Students who can take advantage of such expertise are fortunate indeed.

Whatever the age or the conditions, retirement represents a major transition for most scientists. To the credit of science as a career, withdrawal from active participation, however, gradual, is not a process that is eagerly sought by most professionals. Most of us would like to go on forever, doing just what we're doing.

SUMMARY

Succeeding in science involves a mix of internal satisfactions and external indications of worth. From graduate school to retirement, scientists follow paths which have surprising homogeneity, and only superficial heterogeneity. The game player moving through the continuum of a scientific career will recognize milestones and major decision points—and will approach them with some preparation and insight. Principal decisions involve the first job, the choice of research or teaching, the problem of moving or not moving, and the choice of science versus administration at mid-career. Fortunately, for the well prepared no decision is irrevocable, and intelligent, productive professionals can find challenges and satisfactions at all life stages and at any chronological age.

As in many professions, it is the combination of excellence and chance events that helps determine the route, the rate of progress, and the major accomplishments in a scientific career. Maturation is recognizable in terms of productivity, thorough exploration of hypotheses or concepts, effects of personal or team research on an area of study, and opinions and judgments of peers. Interpersonal relationships are important and often determining factors in advancement, when superimposed on accomplishment in the practice of science.

CHAPTER 8

THE SCIENTIST IN CONTROL
Getting and Using Power

The dependence of those in authority on good science; principal kinds of power in science; some strategies of importance on the road to power.

I tried to point out in the preceding chapter on "Moving Up" that scientists tend to move along a continuum which can be graphed in six major stages. There are places on this continuum where some scientists achieve, use, and sometimes abuse power of various kinds; therefore it seems proper to give some detailed attention to that very interesting phenomenon. We have already considered the transition from scientist to manager, important to many (but not all) scientists. Once the transition is made, accretion of power is a logical concomitant. The science manager assumes a number of legitimate roles, including making decisions about facilities, equipment, and staffing—all the activities that can help ensure a productive, pleasant research environment. He or she is also responsible for the relevance of staff activities, which includes decisions about priorities in research—decisions in which funding is a major consideration.

"Power" as used in this chapter is defined, loosely and probably too simplistically, as the "authority to set goals, objectives, and priorities for yourself and for others, and to determine how and when those goals and objectives are reached." Power is acquired by accident of location or time, by merit and demonstrated competence, by knowing people who make decisions about positions, or by any combination of the foregoing.

There are several truisms about power that could be used to introduce the subject:

- Acquiring power can be at times almost accidental, but *retaining* power over the long term depends on ability and competence.
- In ascending an organizational hierarchy in science, expertise becomes of decreasing importance, being supplanted progressively by managerial-administrative competence and ability to deal effectively with people.
- Holders and users of power in scientific groups must always depend on competent productive scientists, since without good science even the best of other organizational attributes—good fiscal policies and management, effective public relations, good personnel practices—will not suffice. Thus the ultimate power is still in the hands of the scientists, although, as with citizens in a democracy, they at times seem dominated and manipulated by those whom they have designated to lead.
- Most power strategies are designed to manipulate, coerce, guide, or control the satisfactory but unexceptional scientist. The truly outstanding and productive person, though he or she may be caught occasionally in a power game, is usually relatively immune to the manipulations of power players, since exceptional scientists constitute a very critical foundation for the stage on which those with power must play their roles.

The use of power in management of business organizations has been the subject of numerous articles and books. One of the more thorough and analytical was published in 1979 by Kotter.

Defining power as "A measure of a person's potential to get others to do what he or she wants them to do, as well as to avoid being forced by others to do what he or she does not want to do," Kotter based a substantial part of his book on the thesis that use of power is a response to significant *dependence* on the activities of others (in many categories). Noting the decline in "authoritarian" power in modern industrial organizations, he concluded that managers must give major attention to other methods of acquiring power, such as high professional competence, control of communication networks, and encouraging feelings of obligation on the part of others. Abuse of power, according to Kotter, occurs when the power skills of the manager cannot compensate for the extent of dependence on the job performance of others.

Many of the observations and conclusions about power and its uses in business organizations apply, with minor modifications, to most scientific organizations (except, possibly, university science departments). The concept of power as a necessary counterbalance to dependence on performance of others is very clearly illustrated in management of government/industry research groups.

Effective management of scientific research organizations is one of the most complex and demanding of jobs. First of all, good research "directors" are more than just good managers; they must operate from a position of great credibility and continued active participation in science, as well as one which requires good personnel, financial, and other practices. Managers of scientific organizations should be exceptional blends of the knowledgeable, evaluative scientist and the people-oriented, perceptive administrator. Some are and some aren't. Those who are create a pleasant but dynamic environment for productive research. They invest heavily in effective personnel practices, even at some expense to their own personal progress in science. They are interested in the scientific work and professional advancement of their team members. They express appreciation for competence and productivity—verbally as well as in the form of salary and "perks." They insist on continuous development of staff competence through advanced training, both in scientific and in supervisory skills; and they expect excellence. Such director/managers are not abundant, and are to be treasured and supported when they are identified.

There is a special category of science administrators, more properly called "science executives," who function at policy levels concerned with funding strategies, deployment of resources, acquisition of new missions, and commitment of major segments of the staff. These science executives are the real power figures; they are most apparent in government and industrial research organizations.

KINDS OF POWER IN SCIENCE

There are numerous categories of power in science, some of which are unique. Six of these seem significant and worth our consideration here:

1. *Power possessed by the "great one"* in a narrow subdiscipline of science. Such a person does many things each day which impinge on the careers of others, including decisions as a journal editor, opinions about choices for society offices, opinions about hiring, suggestions for symposium participants, and a host of others. To gain some feeling for even a few ramifications of the "great one" concept in operation, read the following case history, drastically abbreviated.

Dr. Sauerwein is an associate professor at a large, eastern university. He is a reasonably bright and perceptive scientist who has in the seven years since his Ph.D. published a number of good papers in an active research field. He has acquired grants, graduate students, technicians, equipment, computer terminals, and expanded office/laboratory space, along with a reduced teaching load. He is clearly a strategist and a gamesman worthy of study, and the sequence of steps in his development during the next five years is predictable:

- *He will spend increasing percentages of his time at conferences, giving invited lectures, and participating in workshops, while his colleagues and graduate students substitute in his courses.*
- *He will be named to the editorial board of at least one journal.*
- *He will continue to publish at least ten papers a year, always with co-authors who have done most or all of the work.*
- *He will edit at least one book a year—usually by organizing a special workshop and volunteering to edit the proceedings, or by asking a number of his colleagues or former graduate students to contribute chapters to a book that he will edit. In either case his name, and only his name, will appear on the cover. The names of the major contributors will be well hidden inside.*
- *He will accept graduate students at the request of selected colleagues, and will recommend his brighter bachelor's degree students to those same colleagues for graduate training.*
- *He will review favorably the books of selected colleagues, and they will review (also favorably) the books he has edited.*
- *He will be elected president of at least one scientific society in his specialty, and will influence favorably the election of colleagues to society offices.*
- *His brighter and more perceptive graduate students will, after getting their Ph.D.'s, move on to other universities to organize their own universe, but will maintain close ties with their former advisor, thereby creating a research network with abundant reciprocities.*

---

2. *Power possessed by the "in group"* in a narrow subdiscipline of science. This is an extension of the "great one" concept to a small-group activity. Members of the group review each other's papers, invite each other to seminars, workshops, and symposia, hire each other's students, plan and control symposia and conferences, and even control journals.

3. *Power possessed by department chairpersons or administrative heads of research groups*, where such positions are appointive and long term. These people make decisions every day that affect the group, and they often have authority to designate directions and levels of research, and to control hiring and firing. There is a wide variation in the exercise of power by such scientific managers, just as there is in any production-oriented enterprise—from dictatorial through paternal (or maternal) to democratic. In university departments there is often a higher degree of participation in major decisions by group members (faculty) than is true in other kinds of research groups.

Usually we think of scientists as fairly strong-minded and protective of their rights as scientists and individuals, but there is a remarkable amount of persistent dictatorship and authoritarianism among the heads of some research organizations. Normally, though, a "winnowing" process occurs whereby those scientists who cannot and will not tolerate authoritarianism migrate to where it is not so apparent, leaving behind a residuum of those who can survive and, oddly enough, even prosper under such management.

4. *Power possessed by the principal administrative officers of granting agencies*, government or private. These people obviously have a large degree of control over the kind and amount of research that is funded. Despite the proliferation of peer review and evaluation procedures, and advisory boards and councils, final decisions about grants are usually administrative, and are made by the hierarchy of the funding organization, with advice from middle-level staff.

To illustrate the routes and dispersion of power in a granting agency, I have selected for an example one of the more recent and (I think) more successful national systems of grants to universities, the Sea Grant Program of the Department of Commerce, just beginning its fourth decade of existence.

Power over the granting process, while ultimately in the hands of the agency administrator (Figure 2) is operationally dispersed to two lower levels within the agency (the grant program director and the grant monitor), one in the university (the grant coordinator/director), and to a lesser extent to agency and university advisory boards. Prerogatives of each level (policy, management, planning, evaluation, strategy) also indicate the extent and nature of power at that level.

What this wiring diagram does not show is that, while power is dispersed, it is still focused at only three operational levels—the grant program director for policy and final funding decisions; the grant monitor for evaluations and recommendations upward to the grant program director and downward to the university; and the university grant coordinator/director for on-site management strategy, planning, and deployment of discretionary funds. Decisions at any or all of these levels can affect an individual grantee, but the primary focus of power is the grant program director.

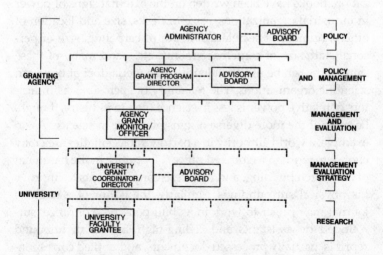

FIGURE 2. Dispersion of power in a grant program.

5. *Power possessed by journal editors.* Publication or its lack can influence scientific careers to a marked degree. The kind of journal in which papers are published is important, as of course is the content of papers and their number. Editorial boards and manuscript reviewers make significant inputs to decisions about publishability of papers submitted, but final decisions still rest with one person—the editor.

6. *Power possessed by scientific peers.* Although their power is diffuse and not always readily visible, scientific peers exert a remarkable degree of control over the careers of their colleagues. Votes of peers control election to society offices; peers examine and comment on research proposals and manuscripts; peers decide whether or not they will cite published work; and peer discussions result in informal but very important evaluations of research.

There are many superficial but not unimportant trappings and symbols of power in science, as in any other human organization. Books have been written on the external signs of power in industrial organizations—size of desks, size and location of offices, number of secretaries, depth of carpeting, size of personal staff, size of conference rooms, etc.—and many of these same signs can be used to identify managerial foci of power in scientific organizations. The reliability of such signs as indicators of relative power is less in science than in industry, though, because of the more diverse origins of power in science. As an example, a world authority in a narrow subject matter area may choose to work in a cluttered office/laboratory in the basement of a research building, and may resist any attempts to improve his physical surroundings. Similarly, the editor of a national journal may prefer to work in a "bull-pen" type of office, surrounded by assistants and ceiling-high shelves of unbound reprints, partially processed documents, and unfiled correspondence. Absence of the more traditional physical facades of power should not confuse the perceptive strategist, however.

POWER STRATEGIES

It is time now to look more specifically at some strategies in the scientific power game.

• *The basic strategy, which is really no strategy at all, is to be very good*—even exceptional—as a scientist. Then many forms of power come more easily, *if they are wanted*—and (surprisingly) many scientists do not want them.

• *Observation of methodologies used by highly successful power players can provide excellent clues and examples,* but the individual scientist must eventually develop his or her own unique set of strategies that depend on individual personality, attitudes, and philosophies.

• *Normally the pathways to power do not lead through positions labeled as "deputies" or "assistants" to key authority people.* Such slots constitute good training for other secondary positions, and they may include reasonable amounts of delegated authority, but they offer no training or practice in the independent exercise of power. The best route to power is through a quantum leap from middle-level supervisor to director. Short-term understudy assignments are alright as long as they don't become habit-forming.

An exception to this rule might be in the case of an assistant to an organization head who is willing—really willing—to delegate full authority and decision-making responsibility in specified broad areas to the assistant. This can be a possible trap, though, once the head person becomes aware of, and possibly threatened by, just how good you really are.

Another possible exception might be as an understudy to a good effective key person who is close to retirement. It is a role that requires great tact, but it can be a learning experience if not too prolonged. The danger, of course, is that the actual retirement may be postponed indefinitely, leaving the understudy with wings dry but with no chance to fly.

• *The power figure develops satellite people who are totally responsive to his or her gravitational pull.* Power figures often have a remarkable ability to make people feel a part of the movement, the action, the inner circle—and once this feeling exists, people will usually respond with maximum effort. The need to belong to the "in-group" which surrounds the power figure can be traced back to kindergarten, to the need for the social acceptance of being part of the class bully's gang. It extends through the teens as a need for membership in the high school "in-group," and to membership in the "in-crowd" of young marrieds.

Response from the satellites is willing, even eager, despite their involvement in more and more activities. Pressure to perform usually results in the satellites' being spread more and more thinly to satisfy the power figure's demands.

• *The power figure is only as good as the bright sycophants which he or she accumulates and controls.* These key people transform ideas into programs, so their careful selection is of critical importance, and their continued loyalty and performance must be assured by a combination of high expectations, rewards, and punishments. Power is ultimately manipulative, and people are the subjects of the manipulation. Manipulation can be based on fear of the consequences of noncompliance or on conviction of the worth of compliance. It may be tempered in the manipulees by desire for recognition or approval, by a need for internal satisfactions to be derived from a job well done, or by a simple need to advance in a particular staff role.

• *The typical power figure needs intelligence, perception, and energy.* Without these characteristics, any reign may be brief. Additional distinguishing marks of some users of power can include common everyday rudeness, pugnacity, and guile. The ultimate power figure for most of us is the dominating, demanding father who is hard to please—but with the entrance of females into positions of power the image becomes a little cloudy.

• *Winners of power games often depend on force of character and brightness—not necessarily on scientific competence or past productivity.* Interplay of political acumen and power in science is increasingly important at higher echelons of government science. The power players must deal with legislators and their aides, investigative committees and their aides, as well as political appointees from within the agency and from agencies with similar missions and interests.

• One of the cornerstones of the highest form of power is the *ability to perceive objectives, and the strategies to meet them, further ahead than almost anyone else.* Perception is not enough, though. It must be accompanied by a visualization of the necessary actions several steps removed from those which directly influence accomplishment of the objectives. This ability separates the superb from the very good in the exercise of power.

• *While many aspects of power acquisition can be learned, there are other ingredients that can only be described as innate or instinctive.* These include a high level of perception of problems and their potential solutions, an extraordinary ability to analyze and synthesize information and to make correct decisions, a basic internal conviction of high self-worth, and a fundamental dissatisfaction with the persistently orthodox.

The bottom line has to be that it is difficult to *learn* how to acquire power, since much of its acquisition depends on innate abilities. It is possible, however, to follow strategies that smooth the path to acquiring power; and it is relatively easy to recognize the uses of power in science, and to understand how such uses produce results.

Political perception and awareness (in the broad sense of the word "political") are important characteristics of a large proportion of scientific power figures. Some ways in which these traits can be recognized are:

a. The person conducts or supervises research relevant to problems of the moment;
b. The person proposes and chairs workshops and symposia which metamorphose into published volumes;
c. The person is invited as an expert to examine and report on scientific problems in other geographic areas;
d. The person maintains a voluminous correspondence;
e. The person serves on agency advisory boards;
f. The person develops a communication network of good scientific research people in other laboratories and in other countries who become friends as well as colleagues.

Often such a person operates from a supervisory or administrative position in a research organization. He or she is able to balance administrative and research commitments admirably because of careful analysis of every developing situation and careful assessment of options. Assessment always includes full consideration of people and their responses.

• *Power is easy to identify, but too elusive for full description.* It can be identified as the consequence of a blend of leadership, managerial style, charisma, motivation, drive, and persuasion. It exists and can be described at three levels—an interpersonal, one-on-one or small-group level, an organizational/managerial level, and a national/international level. Power in science is usually a mixture of levels one and two—interpersonal and organizational.

However they may be dissected and analyzed, power and its uses are realities of organizational life, and scientists are part of and work for organizations—therefore they should have adequate knowledge of the subject. An article in *Harvard Magazine* by R. A. Golde (March/April 1981) states the point well: "power *does* refer to something out there that can be smelt and touched, something immediately recognized when it is seen. Like love, or the functioning of the brain, or the existence of the human spirit, it is real—although possibly unknowable at its core."

ORGANIZATIONAL VARIATIONS IN POWER

It may seem obvious, but it should be emphasized never-theless, that the structure and deployment of power may be substantially different in university departments, government laboratories, and industrial laboratories. Each kind of organi-zation will possess some attributes in common with other orga-nizations and some which are unique.

University departments display a range from genuine peer equality with equal votes—and with a chairperson who is merely an elected or chosen representative—to a "chairper-son for life" structure with that person exerting almost dicta-torial power over kinds of courses and even their content, as well as over selection and promotion of faculty members. The department may have a titular head, usually with a distin-guished scientific reputation, and a managerial head (also a faculty member, but of lesser scientific stature) who is actually responsible for the day-to-day operations. Occasionally a separate institute, with its own head, will function on the periphery of a university department. Its members will have faculty status, but may be paid from grant funds and may do little if any teaching. Other institutes may be autonomous insofar as allegiance to any department is concerned, and may even report to a different level in the university bureaucracy. These complex institutional arrangements provide endless opportunities for strategems concerning laboratory and office space, selection of graduate students, teaching loads, salaries, seminar schedules, symposium sponsorship, and visiting investigators.

Government research laboratories have a number of rea-sonably unique characteristics. Civil service protection pro-duces a high degree of security and long retention, regardless of productivity or its absence. The mediocre may be promoted to positions of authority and power over facilities and people. Rewards for supervisory/management activities usually out-weigh those given for research productivity, so the perceptive often elect the administrative/management route. A common complaint made against government laboratories is that they are foci for "persistent mediocrity" in science, largely because of their mission orientation and the federal promotion system. I have seen mediocrity in all kinds of research laboratories—federal, university, and industry—and I have seen excellent science and excellent scientists in each of these settings. If there is valid criticism of government scientists, it is that excessive amounts of time must be devoted to paperwork in countless categories, and this reduces the time available for scientific production. Another criticism is that good journeyman or mid-level scientists are often siphoned off into sterile staff assignments before they can reach any productive peak in research, and they may never again be heard from scientifically. Executive or managerial power in this environment is often determined by appointment rather than by competence, and appointments can be consequences of previous friendships, political contacts, or favors rendered—criteria which are by no means confined to government research organizations.

In industrial laboratories job security often approaches zero; entire units can be created and destroyed as a result of corporate decisions unrelated to the productivity of the units. Often complete control of the unit is invested in a manager, who may or may not be a good scientist. In other, more democratic units, a matrix form of organization loosens absolute control and disperses power to a number of temporary group leaders. The principles of natural selection—survival of the best adapted, and disappearance of the flawed or infirm—are fully operative in this coliseum. Power moves quickly to the vocal, the perceptive, and the assertive.

The scientific gamesperson will scrutinize the sources and uses of power in any organization which he or she is thinking of joining—to be certain that the existing system is one in which survival and well-being seem possible. Careers are shaped to a remarkable degree by interactions with supervisors, and the system is shaped by a hierarchy of managers to whom supervisors report. Examination of the power structure should be second only to examination of the salary structure in assessing the desirability of a prospective employer.

SUMMARY

Power in science can be subdivided into that derived from *managerial activities* (organizing a research team, directing a laboratory, or serving as an academic dean), and that derived from *professional excellence* (leading a scientific society, editing a journal, or organizing a symposium). Both categories depend for success on skill in interpersonal relationships, and perceptions of correct maneuvers and actions concerning people.

Power is recognizable but not easily definable. Acquisition of *managerial* power may be partly accidental; its retention is not. Acquisition of *professional* power is never accidental; it is earned and its retention is more durable. The existence of this divided base of power in science frees practitioners to some extent from being manipulated or dominated by managerial power figures. External sources of power (recognition by the broader scientific community, receipt of awards, writing technical books) also provide partial insulation.

CHAPTER 9

THE SCIENTIST IN DOUBT
Defining Ethics in Science

Where are the thieves and charlatans; fringe ethical areas; honest scientific squabbling and infighting; maneuvers in ethical combat zones; abuse of data; scientific castration; defrocking phonies.

Of all the controversial subjects considered in this book, probably none has less common agreement than the ethics of scientists. A small and probably unreliable poll of mid-level professionals conducted during a recent cocktail party disclosed no unanimity of views on a "code of ethics," or even agreement that such a code existed or was needed. I reject all these findings.

If the essence of science is a search for reality and truth in the universe, and if the core of science is verification of general laws through observation and experimentation, then the need for an ethical base for scientific activities seems self-evident. This ethical base constitutes a code of practice which should govern the conduct of scientists when they function as professionals. Unfortunately there is a diversity of opinion about the boundaries of some of the principles and practices, and the standards of behavior, which constitute ethical conduct in science. It is precisely this lack of consensus that makes a brief discussion of scientific ethics important to scientific strategists. My operating premise in writing this book is that *nothing about game playing is or should be construed to be unethical, or even of questionable ethics.* In fact game players should be among the strongest supporters of ethical principles as overriding forces in science, because they, more than most others, should understand the need to function professionally within a body of rules.

One surprising (or maybe not so surprising) observation about scientists is the scarcity among their members of genuine charlatans, thieves, and assassins. Most scientists, in their practice of science, are scrupulously ethical. There are a few, however, who hide behind the generalization that "scientists are honest" to invent data, fudge statistics, falsify photographs, even plagiarize the work of others. These are the lunatic fringes of science, and the scientific community takes delight in exposing and denouncing such rare heretics, once guilt is adequately proven. Science is a system of checks and balances, based on the expectation of honesty, that is always weighted on the side of the angels.

This chapter approaches scientific ethics somewhat tangentially. We begin by considering activities at the margins of ethical conduct, then retreat to the semisolid core of acceptable practices, and finally, from the security of this core, we probe a few difficult areas outside the perimeter. Much more could be said about each of these three zones, and nuances of the subject are extensive, but I feel that the basic ethical framework is here, if only in vague outline.

THE PERIMETER OF THE CIRCLE

Although scientists are for the most part totally honest about professional functions, we hear occasionally about questionable activities which must be considered to be on the fringes—close to the edge of the ethical circle if not actually outside it.

One is the *attempted destruction of a scientific reputation* (which is fragile enough at best) by use of innuendo and unproved or unfounded suggestions of misuse or appropriation of data. It is really remarkable how little it takes to contribute to the destruction of a scientific career. Mild hints that all is not as it should be in another's laboratory, the transmission of rumors, the withholding of full approval of research results for unstated reasons—all can add to an unease about a scientist or his work. Often, too, vague hints about dark, unsavory events in a scientist's personal life can have spillover into a career, unfair though that may be. Junior scientists are particularly vulnerable to actions by those wiser in the techniques of infighting. Some devices used are application of that worst of all epithets—"naïve"—to conclusions, or the use of introductions to research papers and review articles to belittle the work of others (usually juniors) who dare to publish in an "authority's" chosen field.

Another (though less common) activity at the ethical fringes is the *assimilation by a senior person of data and conclusions from junior professional staff members and graduate students*—without their expressed willingness to provide the information in the absence of recognition. This practice has so many shades and nuances as to be very difficult to classify. In some institutions which still follow the old classical European tradition in which the director is in a direct supervisory chain to God, it is the expected thing that the director's name will appear on all publications. Fortunately, this tradition is disappearing in the United States (but is not dead by any means). It is the transition from this clearly defined publishing situation to those with increasing amounts of democracy that produces some of the difficulties and unhappinesses about publication of research results.

As an example of the complexity of the situations that can develop, recently several university departments participated in a series of interrelated, jointly planned, and concurrent laboratory studies. Many faculty members participated enthusiastically, and numerous data sets were developed. Some individual scientists prepared and published their findings as individuals or as subgroups. The department chairman at one of the participating universities, who was not personally involved in the experiments or the data analyses but was involved in the planning, prepared and was senior author of a summary article for a prominent journal. The ensuing publicity, even to coverage in weekly news magazines, accrued entirely to that person. The negative reaction within the participating organizations, on the part of scientists who felt that their roles had been unnecessarily diminished, was significant, but not surprising to experienced game players.

Still another relatively uncommon fringe activity concerns acquisition of *advanced degrees*. The awarding of advanced degrees, honorary or otherwise, occasionally nudges the circle of ethical conduct. Usually there is no clearly unethical activity, but some practices can be distasteful to all those who earned advanced degrees through long and serious effort and at the expense of personal comfort and nutrition. The practice of small, relatively unknown institutions awarding degrees to those who are involved in decisions about grants or contracts to that institution is a case in point. There may be some precedent for awarding honorary degrees to benefactors or to distinguished scientific alumni, but any other type of degree granting can have the aroma of "payoff."

A final fringe activity is what we call *"scientific street-walking"* by some university faculty and more private scientific consultants. A cadre of scientists has developed (and is being perpetuated) in many disciplines which makes interpretations of data to fit the needs of those who pay consulting fees. By the consistency of their findings in favor of their employers, it is reasonable to wonder if such scientists have not sold their scientific souls to the devil. Such individuals usually have learned how to speak in terms which do not render them susceptible to refutation in courts of law, or which could be proven to be mis-representations of data. Beyond these expert consultants there is still another questionable group of scientists who work directly as public relations and "front" men for large industries, who are always quoted in public media when crises occur, and who always seem to interpret situations and data in favor of their company. These are the real prostitutes.

Within the circle that would be generally accepted as ethical conduct there is still much playing room for the development of disagreements and hostilities. Novices are often amazed, and unnecessarily disillusioned, to find that heated, sometimes messy, intra- or intergroup squabbles are the *rule* rather than the exception. Only rarely, fortunately, do such affairs come to the surface in the pages of *Science* or the newspapers. Usually they cause only local discord, involve only certain university departments, and may eventually draw in the dean (director) or other university (agency) administrators. Occasionally, though, such unseemly cat fights erupt in public view, and the typical reaction is one of horror to find that scientists are really just people after all, and that as such they can be just as petty and vindictive as any other group of people. The outcome of such scientific squabbles may be departure of one or more of the principals involved; disillusionment of younger scientists, graduate students, and undergraduates; reduction in scientific output while colleagues enjoy the show, take sides, and discuss at great length the pros, cons, and maybes of the situation; and a bit of tarnish on the public image of science and scientists.

Almost invariably, such mini-tempests result from failure to learn or to accept game rules in science. The actions which provoke controversy are usually within ethical boundaries, *at least in the perception of one party to the dispute* (but often not to the other). Except for the activities of the tiny fraction of genuine charlatans already discussed, most friction points concern perceptions of priority of ideas and publication, authorship of papers, claims of abuse of data, promotions and salaries, public expression of aberrant ideas, and other realistically insignificant areas—all of which may be capably handled by the experienced game player, *provided his or her adversary agrees to play by similar rules.*

One such episode resulting in part from failure to understand game rules was publicized at a northern university several years ago. A graduate student, paid from a grant to a faculty member, developed data that were eventually used in the faculty member's Ph.D. dissertation. The student was upset to the point of distributing photocopied handbills outlining his side of the sequence of events and his own perceptions of ethical conduct. The affair came to the attention of the university administration and the outcome was the departure of the faculty member and the student—clearly a no-win situation created in large measure by different perceptions of the scientific ground rules.

It is axiomatic in such traumatic incidents that the *facade of civility be preserved,* that the game continue, but that genuine injury should not go unheeded. This is particularly true when it seems that a relatively new player has been mistreated by an experienced veteran. There are often critical points in a career where a nudge positively or negatively can produce inordinate effects. A casual word of approbation, of little cost or consequence to a senior investigator or professor, can have at just the right moment remarkable influence on the prospects of a bright new Ph.D. An undeserved negative nod at this moment, unconscionable as it may be, can cause delay in or even alter the path of otherwise predictable professional advancement. The ethical routes of access to professional people who inflict unwarranted injury are many, but not always obvious to the novice game player. The junior must be exceptional and productive to play in this game, since otherwise the responses given will seem like sour grapes or the spiteful actions of a scientific upstart.

Scientists in mid-career who are recipients of unnecessary injury should be better able to cope, and should have a larger array of defensive and offensive tools. It is important to the onrush of science that those guilty of gratuitous damage to the career, reputation, peace of mind, or personal satisfactions of another scientist should not go unpunished—but that responses should be within carefully circumscribed ethical limits.

WITHIN THE CIRCLE

It is reasonable, after examining several activities at the margins of ethical conduct, to ask for some kind of codification of what is *within* the circle—of activities that are considered ethical. This can be of great significance to game players, who *always* subscribe to a code that is very carefully ethical. A number of scientific societies and organizations have attempted such a codification, with variable success and with emphasis on aspects that relate to their own particular specialties. Distilling a number of these, I have developed a series of general and vaguely unsatisfying statements about ethical conduct in science. The list is by no means exhaustive, but it may be useful. Some readily debatable dicta are that scientists will:

- Design and conduct investigations in conformity with accepted scientific methods;
- Report in full, on a timely basis, the results of investigations, basing conclusions solely on objective interpretations of available data;
- Not publish or disclose data provided by others without their expressed permission;
- Not publish or release data anonymously;
- Give proper credit for ideas, data, and conclusions of others;
- Prevent release or publication of preliminary or misleading reports of results obtained;

- Resist temptations to utilize news media as first outlets for significant scientific information, in advance of disclosure to peers through normal publication channels;
- Challenge unethical conduct of other scientists, using scientific journals and scientific meetings as proper forums for debate;
- If in private industry, respect the terms of any agreement concerning proprietary information, but avoid entering into agreements which may lead to prolonged suppression of significant new information;
- Provide legitimate conservative estimates of degree of risk of any activity within their area of expertise, based on the best available evidence, but resist pressures for extrapolation and speculation beyond the logical conclusions derived from that data;
- Resist pressures to support decisions based on social, economic, or political considerations by warping conclusions based on scientific evidence;
- Resist pressures to support publicly an officially declared position by an employer if such a position is clearly not in accord with available scientific evidence;
- Offer scientific advice only in areas in which background or experience provides professional competence;
- Resist temptations to express subjective opinions or views in public forums on scientific matters outside areas of individual competence;
- Discourage, by whatever means are available, the employment of professionals in subprofessional jobs, except as temporary expedients;
- Discourage, by whatever means are available, the employment of subprofessionals in professional scientific positions;
- Encourage, by whatever means are available, payment of adequate compensation to professionals for professional services; and

- Encourage the professional development of scientists for whom he or she has supervisory or management responsibility.

My initial reaction to this list is that it is dreadfully dogmatic and incomplete. Additional subjects, which do not lend themselves to easy codification, include participation in union activities, disclosure of information developed for private employers, and public criticism of another professional's work or conclusions. All these are logical ethical concerns, and there are undoubtedly many others which could be included, but I feel that the eighteen items listed above constitute a reasonable core, beyond which the ramifications become almost endless, and are of concern to decreasing numbers of scientists. The core is important to scientists in general and should serve as a modest pocket-guide for evaluating the ethical base for the games that they play.

MANEUVERS IN ETHICAL COMBAT ZONES

Before leaving this important topic of ethics in science, we should probe at least a few subjects in the dimly lit zone outside the ethical perimeter. Of many which might be discussed, I have chosen "Use and Abuse of Data," "Scientific Castrations: A Do-It-Yourself Kit," and "Deflating the Instant Scientist."

Use and Abuse of Data

Much is made of the acquisition of scientific data in a prescribed manner following a statistically valid experimental design, but the real excitement and challenge come from analyses and syntheses of data sets to support or refute existing or new concepts. The use of data developed by others has long been an area of great sensitivity. When does such information become part of the public domain? Is it after initial publication of original data sets, or only after the collector has completed all planned analyses? How long does the collector have priority for analyses before others may move in? What credit needs to be given to the collector when future analyses are made?

Answers to these questions are not simple. They have been debated for decades, and the problems have been exacerbated by computer consumption, digestion, and assimilation of data sets. It seems that the collector must have reasonable lead time to analyze the data. "Reasonable" is a matter of opinion: "reasonable" must be measured in months and not years, and "reasonable" does not preclude development and analyses of parallel data by others. Of course other categories of data—experimental, for example—remain the private preserve of the experimenter until publication, whenever that may occur.

However the boundaries of ethical conduct in data use may be defined, there will be those who disagree. This is an area where careful conservatism may be a useful ground rule, to avoid any trespass on what is perceived by others as private property.

Ideas, data, and conclusions are the principal items of commerce for the scientist. As such they can be subject to a spectrum of manipulations which are clearly or marginally unethical. Included are misuses and abuses of data collected by the individual scientist, or those collected by others. I have attempted to categorize and identify both types. Beginning with data collected by the individual, misuses include:

- *Massaging*—performing extensive transformations or other maneuvers to make inconclusive data appear to be conclusive;
- *Extrapolating*—developing curves based on too few data points, or predicting future trends based on unsupported assumptions about the degree of variability in factors measured;
- *Smoothing*—discarding data points too far removed from expected or mean values;

- *Slanting*—deliberately emphasizing and selecting certain trends in the data, ignoring or discarding others which do not fit the desired or preconceived pattern;
- *Fudging*—creating data points to augment incomplete data sets or observations; and
- *Manufacturing*—creating entire data sets de novo, without benefit of experimentation or observation.

Ideas, data, and conclusions developed by others may be also misused in a number of ways:

- *Premature Disclosure*—reporting, discussing, or citing the work of others which is unpublished or in press, without their stated permission;
- *Scientific Ectoparasitism*—deliberately exploiting or developing ideas or proposals of others, made available in oral or unpublished form for review or comment;
- *Mirror Writing*—utilizing a form of pseudo-plagiarism in which concepts or conclusions developed by others are rephrased or reworded and used without giving adequate credit to sources; and
- *Plagiarism*—outright lifting of data or text from the published work of others without permission from or credit to original sources.

These categories represent only minor expansions of an attempted codification of abuses made one hundred fifty years ago by the British mathematician Babbage, as cited by Hunt (1981). Babbage, in those long-past simpler days, created three descriptors: "forging" (fabricating and reporting results which were never obtained), "trimming" (manipulating data to improve their appearance and utility), and "cooking" (selecting data which fit a hypothesis and rejecting those which do not). How short is the distance we have traveled in all those years!

It is important to keep in mind throughout this discussion of use and abuse of ideas, data, and conclusions, the delicate and often ambiguous matter of *priorities and origins.* Not infrequently, it is difficult to recall or determine the precise origin of a research idea or concept. Furthermore, the same idea or concept may occur to more than one person almost simultaneously. Additionally, observations and experiments may lead separate investigators to similar conclusions.

The struggle for priority and recognition is not new; most of the scientific giants of previous centuries became embroiled in often bitter arguments over credit for original discoveries. The remarkable frequency of simultaneous scientific discoveries and subsequent disputes about priorities has been explored fully in a series of papers by R.K. Merton (1957, 1961, 1963, 1969). Among his conclusions are these: controversies over priority of discovery have been frequent, harsh, and often ugly; controversies often involve even-tempered scientists who seem to rise to heights of indignation only when their perceived priority is threatened; and such controversies are often vigorously prosecuted by friends or associates, rather than by the principals themselves. Merton's general conclusion is that these heated controversies constitute responses of scientists to "violations of the institutional norms of intellectual property."

Game players should take the long view of disputes of this kind, and should be guided by the admonition that any hint or accusation of wrongdoing must be preceded by careful examination of all available facts. The operational principle should be that most scientists are scrupulously honest people, and are innocent until proven guilty. Destruction of a scientific reputation is not an inconsequential act; anything relevant to that process must be approached with extreme caution.

Scientific Castrations: A Do-It-Yourself Kit

Most of the interplay among scientists can be characterized as "competitive but compatible." Occasionally, though, there is an overwhelming need for a less-than-friendly response to a rare unfriendly act. A senior colleague may have belittled your work unfairly in his latest review, a good junior colleague may have been treated unnecessarily harshly by a hostile promotion review committee, a graduate student may have been verbally abused in the discussion following his or her oral presentation—these and similar actions require reactions in the form of minor genital surgical procedures of an ethical kind. Available responses have endless potential variations, but they are all based on a few principles: the essential fragility of a scientific reputation, the ease of slaughter by innuendo, the cardinal sin of naïveté, and the eventual triumph of excellence. The structure of the actions to be taken should be woven around these principles, but must not distort the boundaries of ethical conduct.

Some trustworthy responses—in the case of harsh or unfair treatment in a published paper by another scientist—include a careful, objective letter to the editor of the scientific journal that printed the offensive material, or (in extreme cases) a written request for consideration of the matter by the board of directors of the society which sponsors the journal, or (also in extreme cases) a written request for retraction to be printed in the pages of the journal.

Unfair or biased actions by promotion review committees are more difficult to counter. Often a request for a private session with the committee, a discussion with the dean, or a review by the faculty senate or other oversight group can be effective. Sometimes a recasting of qualifications and accomplishments, combined with a request for reconsideration, can be useful. A final step, if a real problem exists and persists, is the preparation of legal action.

Graduate students are usually intelligent, but are often unprepared for strongly negative responses by faculty members to their oral presentations. If such responses are deserved, then the pain and discomfiture become part of growing up in science; if they are undeserved, then some intrusion by faculty veterans may be justified. Interventions in these instances must not appear to be too defensive or protective; they can take the form of extending the discussion into areas beyond the depth of the harsh reviewer, or of pointing out strongly positive aspects of the presentation. In any event, the situation should be defused, and the self-confidence of the student restored to a degree, by a well-chosen, possibly light, but not frivolous comment. Silence is not acceptable.

Deflating the Instant Scientist

Credibility as a scientist is acquired through intelligence, insights, and productivity. Occasionally, though, there are the impatient ones who aren't interested in paying dues, but want quick rewards. They may be graduate school dropouts, technicians who have learned the jargon but little else, or outright phonies. Harsh treatment is the only course of action once these anomalous individuals have been identified.

A classic case of the "instant scientist" was related to me by a colleague from a large, western university. A middle-aged, impeccably groomed man in a foreign naval officer's uniform appeared one day at the weekly "brown bag" seminar for graduate students. He had letters of introduction to the professor in charge from several well-known European scientists, and was introduced to the group as an expert on bird migrations. He presented an admirable series of seminars over the next several weeks on the subject of long-distance migrations, built around advanced and complex measurements of the earth's electromagnetic fluxes. He was an extremely gregarious type and was readily accepted by the group. Then, suddenly, he disappeared! The professor soon began receiving a flood of hotel, restaurant, and other bills, which the pseudo-scientist had conned unsuspecting neighborhood proprietors into accepting. A check of his letters of introduction disclosed that he was a complete fraud; a check of the naval service confirmed this. Even his so-called reprints were fakes, and were not actually published in the obscure journals that were given as sources. The awful and distressing thing was that all of the participants, including the professor,

were taken in completely by this brilliant scam. The only saving fea-
ture of the entire episode was that the impostor had developed a friend-
ship with one of the female graduate students, and when he wrote from
a new address she squealed on him—enabling the professor to recoup
at least part of the expenses of the bills which had accumulated and to
expose the fraud at his new location.

Fortunately, such episodes are rare; many scientists will
survive an entire career without encountering outright
phonies—but they are out there, ready to take advantage of the
basic trusting nature of most of us.

Another example of the relish with which the scientific community
exposes phonies was given good coverage in the pages of several
national journals more than a decade ago. A person with some med-
ical training, and presumably with a medical degree, moved rapidly
for three years through the United States academic establishment,
gaining an honorary M.D., publishing over sixty papers, doing can-
cer research at several institutions, joining numerous scientific soci-
eties, and driving a yellow Cadillac. Everything was based on fraud.
He used the background section of a United States researcher's grant
application as a review paper, publishing it word-for-word under his
own name in three different journals. He purloined a manuscript sent
to a colleague for review, put his own name and those of two fictitious
co-authors on it, and sent it off for publication in a foreign journal. He
took, almost verbatim, a paper published by others in a Japanese jour-
nal, put his own name on it, and published it in a European journal.
These almost unbelievable exploits are still causing reverberations in
journals which published his papers (some sixty in all), and undoubt-
edly some private soul-searching by editors who were duped. The cul-
prit has disappeared, so his motivations are still matters of conjecture.

These instances are interesting and bizarre, and good scientists hesitate to get involved, yet there comes a point where
someone must, and usually does, yell "fake."

SUMMARY

The act of doing good science is surrounded by an intangible but still very real envelope of ethical behavior. Many attempts have been made to codify its components—or at least to define the ephemeral boundary between that which is ethical and that which is not. Strategists function carefully and consistently within the envelope. Some probe its margins gently on occasion and find them hot and painful. Rarely, a tiny minority of scientists elects to move outside, to ignore the boundary deliberately, in search of gold or some other worldly prize. The essential conservatism of science dictates that aberrant behavior of this kind should result in eventual penalties, even though culprits often escape for prolonged periods.

PART THREE

SPECIAL INTEREST AREAS FOR SCIENTIFIC STRATEGISTS

In Part Two of this book we considered some of the critical issues in scientific game playing—transitions, power, and ethics—but there is still much to be discussed in specialized strategic areas, such as how women and men relate in science, how scientists interact with bureaucrats, how external forces such as news media and the public impinge on science, and the uneasy alliances which may exist between scientists and industry.

Part Three takes these topics for a small ride. Each treatment has great potential for expansion; each subject can influence scientific careers, singly or multifactorially. Unlike the topics in Part Two, which are each of overriding significance, the topics in Part Three are more important as background to success and pleasure in science. This does not reduce their value though, since it is the total mix of strategies that can separate the superb from the merely very good.

The emphasis here, as in earlier parts of this book, is on correct interpersonal activities. The playing field is, however, expanding in some of the chapters beyond the laboratory, classroom, and conference room, to a peripheral zone where new strategies can be important. We begin with the changing role of women in science, then move prudently to a mixed bag of interactions with non-scientists, includ-

ing bureaucrats, reporters, lawyers, citizens, politicians, and business managers. The final chapter has a more commercial focus, examining the scientist in industrial research and the scientist as a consultant.

CHAPTER 10

WOMEN IN SCIENCE:
A CURRENT APPRAISAL

The movement toward equality: current status of women in science; impediments to achieving full equality for women scientists; women scientists in the corridors of power; the gender gap at the apex.

The title of a recent book, "Why So Slow: The Advancement of Women," by Dr. Virginia Valian of Hunter College (1998), asks a question pertinent to the careers of women in science. "Why So Slow" expresses a legitimate concern about movement toward equality by many (but not all) female professionals. Others—mostly younger faculty members—find few if any gender-induced differences in their academic experiences. Although *velocity* may be in question, as it is in Dr. Valian's book, substantial progress has been made since the 1970s in leveling the professional playing field for women scientists. Some gullies and hummocks persist, but more and more women perceive a barrier-free landscape outside their office windows.

Reality, it seems, is in the eye of the beholder—male versus female, recent Ph.D. versus aging female assistant professor, male laboratory director versus female program leader, female laboratory director versus male program leader—all trying to understand and adapt to a changing interpersonal climate in that most conservative of human inventions: science. The bad old days of overt discrimination against women scientists are gone, but enough subtle remnants exist, even today, to cloud the careers of many women, and to make their lives in science marginal instead of the deeply satisfying experiences that they should be. How close are we to a kind of community of professionals that judges individuals on the basis of competence and productivity, not on gender? I think we are closer than some would have us believe, but that is obviously a male perspective.

THE MOVEMENT TOWARD EQUALITY: CURRENT STATUS OF WOMEN IN SCIENCE

There is no question that obvious discrimination against women scientists in hiring, advancement, salary, and recognition existed well into the 1970s. Women up to that time had been relatively scarce in science, except as technicians or glassware washers. Today's new women scientists enter a world that is radically different from that encountered by their predecessors. The feminist movement and federal anti-discrimination legislation encouraged damaging assaults on science as a predominantly male bastion. Three decades of action against overt and subtle discrimination and stereotyping of female scientists have had fantastic effects on the status of women in science—to the extent that many entry-level female scientists fail to see any gender-related problems at all. They may have heard woeful tales from older colleagues about the time when "good old boys" clubs dominated every aspect of science and scientific organizations, and women were considered to be poorly qualified interlopers, but they feel little discrimination themselves.

I was a naïve bystander, witnessing with lack of any real comprehension the daily acts of discrimination that were part of the professional life of Joan R. Traverse, Ph.D., who for twenty years had been an assistant professor at a northeastern state college. From my fuzzy perspective as a pushy undergraduate zoology major, she was an intelligent informed teacher, who also did research, but under difficult circumstances. She was responsible for several large undergraduate courses including all the labs and the preparation for them. She had no graduate or undergraduate assistants—unlike the men teaching other core courses. She was also the only female member of the department, which numbered about fifteen faculty people.

Those were the exciting days just after World War II, when hordes of men in their early twenties, recently released from military servitude, flooded college campuses. They demanded the best efforts from teachers and did not suffer fools or pretenders lightly. Many of these veterans instinctively recognized the quality of Dr. Traverse's teaching and the anomaly of her position. Her courses were sought by those looking for substance, but she remained an assistant professor for the rest of her career.

My thought at the time was (and still is) "What an inexcusable travesty, in a society desperate for exceptional, inspiring teachers." But this was in the early 1950s, before the visibility of overt discrimination against female scientists became high enough to warrant attention, and attitudes began to shift. No action was even considered then, either by students or by the college administration. We sat quietly as Dr. Traverse and other female professionals caught in the system suffered silently.

But those bad old days are rapidly disappearing from living memory. They have been replaced, slowly and incrementally, by a near-egalitarian community—still imperfect in many ways—but almost free of gender-induced hindrances to success. At present, most of the overt forms of discrimination have disappeared, or in some instances have gone far underground.

Women populate the halls of science in ever-increasing numbers, although they are still underrepresented at higher academic ranks and in some of the inner circles of science.

One plausible reason for the changes that have resulted in the present state of near-equality of the sexes in science practice seems to have been the increasing *numbers* of professionals in every kind of scientific endeavor. Where a token woman might have been a member of a research group in 1970, today women may outnumber men, or are at least proportionally well-represented in almost every organization.

Numbers alone can have a salutary influence on acceptance of female professionals. The sole woman scientist in a group of male colleagues, or a tiny minority of women scientists in that same group of males, will be seen principally in terms of gender—but when women begin to appear in greater abundance in the professional science workplace, men do not continue to view them only in terms of their sex, but begin making judgments based on their competence. Several studies have indicated that when the number of females in the workforce—any workforce—reaches some significant level (social psychologists estimate about 15 to 20 percent), then a so-called "tipping point" is reached when women are no longer scrutinized as specimens because of their gender, but begin to be evaluated in terms of ability and contributions.

The untrue but persistent stereotype that "women are less technically competent" also seems to fade when a number of women are present in the group and the work atmosphere changes in this way. A "sea change" in attitudes takes place on scientific research teams, university science faculties, and in cohorts of graduate students when more women appear and their presence in significant numbers helps to reduce tensions and to raise the comfort level in these and other groups that are assembled for specific purposes. As was pointed out in a recent *New York Times* feature story by Cornelia Dean (New York Times, Science, November 10, 1998, pp. 1 and 4), at the tipping point, "A more civil culture develops, [in which] the women begin to censor their own behavior, and the men keep other men in line." The concept undoubtedly has complexities known only to social psychologists, but even simply put it seems to explain behavioral changes in mixed gender scientific groups. More importantly, it agrees with observed changes in group behavior as a research team expands with time and funding, and the ratio of females to males increases.

It should be instructive at this point to look beyond generalities at a more factual assessment of the status of women in science, as measured by the important criteria of salary, position and recognition.

SALARY

Pay scales are of course very private matters to most scientists, so they have to be investigated cautiously, and the results interpreted conservatively. A study in the early 1970s by the National Education Association (NEA, 1972) disclosed that salaries of women scientists were less than those for men by only 5 percent at the instructor level, 6 percent at the assistant professor level, and 10 percent at the full professional level. More recent surveys, including that of Yentsch and Sindermann (1992) have suggested that the small salary gap has narrowed still further, to relatively insignificant proportions, except at the highest professional levels, where there is still a significant difference between salaries of males and females. In one of those recent surveys, a majority (75 percent) of the women scientists queried estimated that their salaries were equal to or approximately equal to those of male counterparts at the same institution at the same career stage and with similar capabilities. Only one woman stated that her salary was higher than her male counterparts, but almost 20 percent said their salaries were slightly lower than male counterparts.

Other tentative conclusions about salaries were:
- Financial rewards have become as important to women as to men, but to most women the financial rewards are often viewed as a criterion of recognition by others of their value to science, rather than as a career objective.

(This is not true, of course, for single mothers or for women who are the principal wage earners in the family).

- Financial rewards are often viewed by women as an important component of "acceptance by the system"— which includes institutional parity with male scientists in rank, salary, and tenure.

POSITION

A recent symposium on women in science disclosed remarkable gender differences in academic rank. Women predominated over men at the assistant professor level (31 percent to 17 percent), but then the ratios were reversed at the higher ranks: 50 percent men to 38 percent women at the associate professor level, and 17 percent men to 9 percent women at the full professor level, indicating a lower promotion rate.

Other tentative conclusions were:

- Promotions in rank still do not reflect gender parity, and many women, because of their more recent entry into science in numbers, remain clustered in the lower ranks.
- Traditionally, women who succeed in science are said to do so because they mimic men, with behavior patterns that include aggressive self-confidence despite loss, failure, or defeat, and the ability to "walk the earth as if they owned it." (Many women do not buy this as a work ethic, substituting a gentler, more sharing approach).

RECOGNITION

Recognition in science has a dual nature. *Institutional* recognition is usually in the form of advancement in rank and salary, and *professional* recognition from the larger scientific community is usually in the form of invitations to participate in conferences and symposia, award of prizes, and election to society offices. Institutional recognition (rank and salary) has, as we have just seen, a diminishing but still existing gender component, especially for advancement to the higher academic ranks, but professional recognition is almost free of gender considerations. The only persistent inequalities are the exclusionary in-groups and male cliques or "good old boys" clubs that still characterize some narrow specialties—and even these are becoming more and more obsolete.

So, a cursory examination of basic criteria of gender equality, as reflected in salary, rank, and recognition, leads to the conclusion that inequities have been reduced markedly since the 1970s.

IMPEDIMENTS TO ACHIEVING FULL EQUALITY FOR WOMEN SCIENTISTS

Despite the progress made in the past several decades, persistent gender-related inequities still exist in how science is practiced and managed. Salaries for women are not always comparable to those for men; promotions for women lag behind those for men in the higher academic ranks; and men gain acceptance into the inner circles of science more easily. Certainly these imbalances have diminished in many institutions and have virtually disappeared in others, but roadblocks still exist on the route to full equality for women scientists. Some of the reasons that have been suggested to explain the differentials that remain are these:

- *Male scientists publish more technical papers than female scientists do.* More than fifty studies have supported this finding (Cole and Zuckerman, 1984). The disparity continues, and even increases throughout scientists' careers.

Many reasons have been proposed to account for this imbalance in productivity (Cole and Singer, 1991). The most obvious is sex discrimination—certainly important in pre–1970s science, but probably with less impact on newer generations. Other possible reasons for lower productivity of women include lower motivation because of early cultural barriers; reinforcement—positive for males, and negative for females—in the form of rewards and recognition for early productivity; cumulative advantage created by men's greater access to scientific resources because they produce more early on; domestic responsibilities of women (child bearing and child rearing in particular) that place some women at a competitive disadvantage when career productivity is important; and organizational factors (such as teaching in a small college with a heavy course load) that discourage research and publication. A final theory to explain why publication rates of women are lower than men was proposed by Cole and Singer (1991) as "limited differences" in which individuals are exposed to a series of major and minor events whose outcomes, if negative—such as losing the vote for a society office or not receiving a grant—accumulate and reinforce responses to other events. Each event and each reaction is limited in its long-term effect, but the *cumulative* effect of these small differences can result in imbalances in career productivity between men and women.

- *Female scientists may have had mentoring experiences in graduate school that are different from those of male cohorts.* Mentors are responsible for imprinting on their students the essential elements of science—its history, traditions, ethics, values, approaches to thinking, analytical methods, evaluation criteria, and publication guidelines—all factors that are important to achieving success. In addition, mentors act as "sponsors," who ensure that their students are introduced to and participate in the *collegiality* of the scientific community. Female students who have male mentors frequently find that this important element is missing from the mentoring relationship, possibly because the mentor is afraid of how others will view his interactions with them, and will avoid the informal exchanges and subsequent acculturation that occurs during the closer associations with male students and male mentors. Such deprivation can be very disadvantageous to the beginning careers of female scientists.

Most male faculty members deny any gender considerations in accepting graduate students or in their relationships with them—as well they might in today's academic environment favoring reduction of any subtle discriminatory practices. Similarly, female faculty members deny any gender-based selection, but most of their students are often female.

- *Female scientists with children may suffer from discontinuities in their careers,* or at least overwhelming stresses associated with child bearing and child rearing, that can impact negatively on productivity. The effects of those discontinuities—dropping out temporarily for family reasons or functioning only part-time as a scientist—can slow progression in science, and can even be viewed as lack of full commitment to professionalism. Some studies have suggested that female scientists with children are just as productive as those who are unmarried, but too little is made of the stresses that must accompany those achievements.

- *Female scientists may be undervalued by some male colleagues.* Studies also show that male scientists, possibly for reasons of tradition or custom, are sometimes perceived as slightly more competent and more qualified than they are, and female scientists are considered slightly less competent and less qualified than they are. This imbalance, when added to many others—advantages to males in promotions, continuing disparities in salaries at some institutions, occasional exclusionary acts by male col-

leagues, failure of male colleagues to support a reasonable position proposed by a female faculty member—all
contribute to what sociologists call "undervaluation" of
female scientists. These factors are not usually overt, and
are not necessarily designed to be hurtful or discriminatory, but they can contribute to feelings of inadequacy
that affect women scientists' careers.

- *Intellectual risk-taking is often considered a characteristic of*
  male rather than female scientists. Men, for example, seem
  more inclined to publish a tentative hypothesis with limited data than women, who may believe that they are
  ". . . less likely to be given the benefit of the doubt, and
  their proposals [more] easily dismissed" (Valian, 1998).
  This attitude, according to Dr. Valian, may help to
  account for the lower production of technical papers by
  female scientists, if they assume that their research
  results may not be evaluated by the same criteria used for
  the work of male scientists.

So, in summary, despite great progress in eliminating overt
discrimination and reducing gender stereotyping, impediments to full equality of female scientists still exist in the form
of small imbalances and residual biases. The problem is that any
small imbalance, such as the one favoring males for promotion,
can, over time, result in perpetuating male numerical domi-
nance at the higher professional ranks. Furthermore, these
imbalances in the treatment of males and females can accumulate and perpetuate the disparities that still exist.

WOMEN SCIENTISTS IN THE CORRIDORS OF POWER

With the appearance of more and more female scientists in
research laboratories and other science-related organizations,
there has been an accompanying increase in their presence in
supervisory/managerial positions. More women today are pro-

ject leaders, program managers, mid- and upper-level bureaucrats, and executives of science-based companies. Women are making decisions about funding, recruiting, deployment of resources, promotions, research objectives, and many other activities critical to the well-being of the science establishment.

The increase in women in positions of power and influence is a recent phenomenon, becoming most apparent in the 1980s and especially in the 1990s. Those positions had previously been occupied almost exclusively by men; women's access had been limited earlier by gender differences in training, encouragement, and financial support. All the imbalances have not been corrected, though. Persistent limitations to movement into managerial levels of science include these:

- More than half the women in science have fewer than ten years of work experience as professionals, as compared with one-fifth of the men. This translates into a smaller applicant pool of females than males qualified even to be considered for senior level supervisory/management positions.
- Another possible limitation women face in getting key supervisory/management jobs is that the final selection is still controlled principally by men.
- A further limitation is that, in some organizations, advancement of women to supervisory roles that are within reach of management level jobs is slower than it is for men, as reflected in differences in salaries and job descriptions.
- Women may be blocked from reaching the choice top-level management positions by the existence of a so-called "glass ceiling" or "invisible ceiling"—the subtle barriers that develop from obsolete stereotypes and unsubstantiated assumptions (such as "women are too emotional to be good managers," or "a woman will let family responsibilities interfere with her work," or "women do not have the necessary knowledge and expe-

rience"). Some pessimistic viewers of this state of affairs have expanded the metaphor, aptly describing women as "being trapped between a sticky floor and a glass ceiling."

- Mistaken *perceptions* of women scientists in managerial positions, if not corrected forcefully, can retard or prevent advancement to key jobs. Some of the worst are:
  — "Women are tokens of equal employment policies that dictate appointment of females to various supervisory/management levels regardless of competence."
  — "Female managers are too concerned about the personal needs of employees and cannot make hard decisions that have negative impacts on staff members."
  — "Female managers avoid confrontations, or react harshly or irrationally."

These and other misperceptions can and do retard or prevent women from moving more freely into the managerial levels of science, but their experience in science management is not as bleak now as this discussion might suggest. There is more openness today in scientific organizations about gender parity in opportunities for those with equal abilities and background. The newer generations of male scientists have been less inclined to perpetuate female stereotyping and more inclined to make judgments about peers based on ability, intelligence and productivity. Newer generations of female scientists, on the other hand, are less inclined to accept even the slightest degree of inequality based on gender; in fact most of them have entered science without any fear of bias or prejudice. They fully expect to be scrutinized and evaluated as professionals, regardless of gender—and in the present academic environment they usually are.

The Gender Gap at the Apex

The topic of "The Scientist as a Manager" was explored earlier; that discussion was almost gender-free, as any proper introductory discourse should always be. But now we have come to a chapter titled "Women in Science," in which a generally optimistic view of the current status of female scientists has been stoutly maintained—even when the data are not always especially robust. One disquieting finding is that women are scarce at the higher levels of both academic and governmental science organizations. In one of the federal bureaucracies, according to a recent independent study, women occupied only 7.5 *percent* of the most demanding agency positions (Clemmet, 1990). Explanations offered for this disparity resurrected most of the old stereotypes and a few additional ones: "women are not willing to commit to the long hours required," "the presence of women may antagonize constituents in technical fields, who expect to interact with a man," and "male politicians are more comfortable talking to a male scientist." A thoughtful analysis resulted in a more rational list of possible explanations:

- Women scientists are entering administrative and bureaucratic jobs in larger and larger numbers, but the persistent imbalance in proportions of males and females and the time it takes to climb the hierarchical ladder means that the female candidate pool for the apex positions is still small, when compared with that of males.
- Women scientists who qualify as candidates for key positions have to struggle hard for increased visibility and broadened contacts. They must involve themselves deliberately in confidence-building public interactions and professional presentations if they are to emerge from the shadows of their male counterparts who are competing for the positions.
- Mere scientific credibility is not sufficient in the rarified atmosphere of top level jobs in science. Other kinds of expertise, extending far beyond science into areas such as politics, negotiation, diplomacy, and economics, may be significant arbiters of success.

- Some achieving female managers become afflicted with the so-called "Impostor Syndrome," described as feeling that they are frauds, lacking in the ability, confidence, and self-esteem needed to climb further. The syndrome may be caused in part by early imprinting or by reinforcement of feelings of inadequacy, engendered by stresses imposed by "the system." If not confronted, this syndrome may cause derailment of a promising career.
- One final explanation for a gender gap at the top is *differential attrition;* women disappear from the pipeline at every phase of their careers, beginning in graduate school. By the time a given female cohort is ready for the choice senior positions, most of the group has abandoned science altogether. Men also disappear from science, for a variety of reasons, but not in the same relative numbers.

So the gender gap in the highest science positions—deans and provosts, agency directors, heads of granting foundations, and executives of high technology enterprises—is real and it is continuing. Overt discrimination does not seem to be a factor, even though subtle forms of belittlement and isolation may persist in some organizations. We are left, then, with a number of other possible explanations for the disparity; one of the most logical is the differential departure of women from science at every career stage, leaving at the end of the pipeline a drastically diminished pool of potential candidates for key science positions.

SUMMARY

The primary career goals of women scientists include (but are not limited to) (1) contribution of significant new information based on productive research, (2) acquisition of recognition and respect from peers and colleagues, and (3) acceptance as equal participants by the still male-dominated system of science. Salaries and promotions remain as institutional problems impeding full parity of women with men, although these problems are shrinking as the numbers of women professionals increase. Gender parity has been achieved in many academic institutions, but, in some, pay scales for women scientists are still lower than those for men—even though the gap has narrowed in all but the highest ranks. Gender parity in rates of promotion has not been achieved in many colleges and universities; women still cluster in the lower academic ranks.

A combination of the early feminist movement with the passage and enforcement of equal employment legislation aided a gradual improvement in attitudes and practices of the scientific establishment (male) as they concerned female professionals. Continued progress, at a rate that encouraged some observers but discouraged others, characterized the 1980s and 1990s, as more and more women occupied academic, governmental, and industrial positions in research and other science-related fields.

Despite some gender-related inequities at the institutional level, only scattered remnants of bias can be found in the larger scientific community. Professional recognition in the form of awards, prizes, and election to society offices is now remarkably gender free (with certain exceptions discussed earlier), and is based largely on merit and performance. One persistent roadblock to equal professional recognition is the puzzling disparity in scientific productivity between male and female scientists; males publish significantly more papers during their careers. A number of plausible explanations have been proposed and discussed here, but the issue remains unresolved. One aspect of this problem seems clear at present: women who do publish extensively have better access than ever before to the higher faculty ranks at major universities. With the entry of younger professional women in greater numbers, the proportions of the productivity question should diminish—as most of the components of bias have during the past several decades.

CHAPTER 11

COPING WITH BUREAUCRACY
AND BUREAUCRATS

An attempt to describe bureaucrats and to define bureaucracy—types of bureaucrats and kinds of bureaucracies: the politician-bureaucrat, the career bureaucrat, and the time-server bureaucrat; the government science management bureaucracy, the scientific administrative bureaucracy, and the bureaucracy created to disburse research funds; the inevitability of contact with bureaucracy and bureaucrats; federal grants as the principal scientific interface with bureaucracy; grant review procedures—the site visit in particular.

The words "bureaucrat" and "bureaucracy" have a variety of connotations, often negative, depending on one's previous exposure. In a brief survey made recently, principal responses to the word "bureaucracy" included "self-serving," "immoveable," "wasteful," "impersonal," "conservative," and "unyielding"; principal responses to the word "bureaucrat" included "self-serving," "self-protecting," "unwilling to make decisions," and "time-server." Only rarely during the survey were favorable comments elicited from respondents.

Dictionary definitions of bureaucracy and bureaucrat carry equally negative connotations. Bureaucracies are defined as "administrations characterized by diffusion of authority and adherence to inflexible and complex rules of operation which impede effective action." Bureaucrats inherit the derogatory connotations of this definition as "officials who insist on rigid adherence to rules, forms, and routines."

Bureaucracies are usually associated with some kind of governmental function, even though they exist in every organization, public or private. Much of the discussion in this chapter concentrates on government bureaucracies, because of their particular current relevance to science and their pervasiveness in today's affairs.

To explore even the fringes of the fascinating but complex bureaucratic universe, some guidelines and descriptions might be useful. It is important at the outset to discuss *bureaucrats* separately from *bureaucracies*, since there are many kinds of bureaucrats and bureaucracies—supplying the ingredients for a complicated and totally confusing matrix. We should logically begin with the bureaucrats, since people are the keys to organizations.

TYPES OF BUREAUCRATS

Bureaucracies are peopled with distinct and only weakly intergrading types of bureaucrats who have been variously categorized, but are here described as (1) the politician-bureaucrat, (2) the career bureaucrat, and (3) the time-server bureaucrat. Subtypes exist, of course, but it is quite likely that when you encounter a bureaucrat he or she will fit reasonably neatly into one of the above pigeonholes.

Because people of these three types constitute the essence of bureaucracies (along with various administrative, clerical, and support personnel), it seems important early in this discussion to describe them as specifically as possible. Scientific game players should develop the ability to identify each type after no more than seventeen seconds of conversation, since the proper approach and the proper words to address each type should be quite distinct.

The Politician-Bureaucrat

A former Pentagon official, Leslie Gelb, quoted in an article in the *New York Times*, made a workable, if incomplete, distinction between those whom he termed "bureaucrat-politicians" and "bureaucrat lifers" (Shenker, 1972). Bureaucrat-politicians, according to Gelb, maintain a studied neutrality on issues, preferring to listen to evidence and argument and to play the role of mediators. They are also good actors, with rewards expected in terms of deference, being consulted, favorable press reaction, and internal satisfaction in doing something right. Bureaucrat-politicians quickly master bureaucratic jargon with its varied connotations. They enjoy power, but recognize the limitations of power at any level of bureaucracy. (Bureaucrat-lifers, on the other hand, are "nine to five careerists who take all their sick leave and vacation time, patiently learn their jobs, and the unwritten rules, and rest cautiously content with moderate ambitions"). Scientific game players with any expected contacts with politician-bureaucrats would enjoy and profit from Gelb's comments in the above-cited article, retrievable, I am sure, from the publisher's morgue (March 5, 1972), and still relevant three decades later.

There are other perspectives on politician-bureaucrats that can be useful and that extend Gelb's concepts. Probably the most important is that they must depend for much of their success on "career bureaucrats" (to be described next). The delicate accommodation, at federal and state levels, of political appointees and career civil service managers can determine the success or failure of programs. The sensible politician-bureaucrat will always have an experienced, outspoken, system-conscious career bureaucrat as part of the inner circle—and will listen to that person's opinion in most matters, as partial insurance against making serious mistakes. This person provides a very necessary contribution of "institutional memory" to any discussion, when such an input is relevant. The association of political appointee and old-line bureaucrat can be fragile and easily disrupted, though, since it represents a temporary working relationship between inhabitants of two different worlds. The association has survival value for both parties; expediency is its sole reason for existence.

Policy-level political people must rely on career civil servants to translate plans into programs. The critical importance of this interaction was the subject of a perceptive book by Hugh Heclo (1977) titled *A Government of Strangers.* Politician-bureaucrats are transients, and as such, poor ones can be waited out. Good ones, on the other hand, can bring vigor, fresh ideas, and new approaches to line agencies, if they proceed with good judgment and reasonable restraint, and act consistently within the policies and beliefs of those responsible for their appointments.

Politician-bureaucrats have and exercise power during their tenure, but it is the reflected power of the elected official at the peak of the hierarchy, and is not necessarily or usually derived from personal competence of the bureaucrat. The proper use of this power is to bring agency policies in line with those enunciated by the elected official; the power is often misused.

Effective politician-bureaucrats can perform a very critical function, as the connecting link between the existing bureaucracy and elected politicians, whether they are in the administration or in the legislature. This is really the only *essential* function of a politician-bureaucrat, even though incumbents constantly try to parlay the job into other agency activities which are better handled by career bureaucrats. Principal activities include briefings of congressional subcommittees and aides, discussions with legislators and their aides on issues associated with proposed bills, assistance to legislators in drafting bills, and explanation of agency policies and positions on particular bills and acts.

Additionally, politician-bureaucrats interact constantly with their politically appointed peers and counterparts in other agencies—in interagency policy and coordinating committees, in joint briefings for agency heads, and in crisis-response sessions.

The politician-bureaucrat thus has a reason for existence, provided that he or she recognizes the proper role to be played, is intelligent and perceptive about playing that role, and occasionally asks for advice and information from career bureaucrats.

Because of the tenuous tenure of political bureaucrats, they have developed unique survival techniques. Policy levels expect to be replaced with a change of administration, regardless of their effectiveness on the job. They will spend a significant part of the "sunset months" of their appointments in search of suitable berths outside the government—in private industry, foundations, teaching, or consulting—often (but not always) to reappear in important government jobs with a subsequent change in administration. Intermediate levels of political bureaucrats secretly hope to survive changes in administration, but frequently find themselves on the street also. During their brief moment of power and influence, political bureaucrats can make substantial impacts on the agency, depending on their skills and on the extent to which they use the career bureaucrat cadre wisely. Often, however, effects are short lived since each succeeding political appointee has his or her own mark to make; this requires repudiation or drastic revision of the key operating programs of the previous appointee.

Each appointee feels bound to create or impose a distinctive management style or a program with a catchy title or acronym. Political appointees during their reign can also affect the lives and fortunes of several intermediate levels of career bureaucrats beneath them—by promoting or failing to promote, by judicious selection of assignments, by encouraging or suppressing visibility, or by deliberate shelving (with no loss in pay) of career people who formerly occupied roles of importance and who may not seem compatible with the rest of the political appointee's new team.

An almost classical example of the power available to the political appointee was a case in which the appointee was named as director of a relatively small research agency of one of the federal executive departments. One of his first official acts was to replace a career bureaucrat who had served with distinction as chief of a principal division of that agency—banishing him to a small office with a poorly defined mission in a remote location (with no loss of pay or grade, but with a drastic loss in prestige, power, and ego). It was later disclosed that this good career bureaucrat had, years earlier, selected another candidate for a job over the person who was now his boss (the political appointee).

The career man never returned to a national position, even after the political appointee disappeared (predictably) with a change in administration—possibly, in part, as a consequence of the trauma resulting from his (the career bureaucrat's) removal from a key position because of political maneuvers.

The game player might be quick to point out that episodes similar to the above are commonplace, and should be accepted facts of life for those who rise to positions of power—that the risk taking is always directly proportional to the level of the position. A career bureaucrat—even a good one—can, however, have trouble recovering from such banishment, since the number of positions near the peak of the hierarchy is limited.

The Career Bureaucrat

Government departments and agencies function because of the efforts and expertise of career bureaucrats. Continuity and progress in government programs are assured by career bureaucrats.

Some characteristics of good career bureaucrats include the following:

- They are good team players, capable of shifting loyalties and accommodating gracefully while still remaining faithful to basic principles.
- They are a principal source of stability, expertise, continuity, and institutional memory in any agency or organization; they are the dominant force for "organizational homeostasis."

- They are generous in the personal time invested in their job, and are fully involved in and committed to it.
- They tend to avoid strong initial advocacy positions on many issues, preferring to see the data and evidence first, and to leaven this with any official agency position. They are careful to avoid extreme or individualistic positions on most subjects.
- They expect rewards and promotions commensurate with their performance and productivity.
- They are capable of a remarkable degree of self-hypnosis—to convince themselves of the worth and value of any assigned task.
- They tend to be people with enthusiasm and loyalty. They also tend to possess good analytic and synthetic abilities.

Looking specifically at *scientific* career bureaucrats, there are interesting subspecies to be discerned and classified. As a beginning to a taxonomic key, we offer the following:

- *Antepenultimate Scientific Bureaucrat:* can talk with smoothness and much jargon, but says absolutely nothing; makes no commitments and gives no clear indication of his position on any issue; departs restaurant small-group meetings early to avoid paying the bill.
- *Penultimate Scientific Bureaucrat:* enjoys simplified conceptual scientific discussions, but becomes withdrawn and hostile when specifics are introduced; likes broad and untenable generalizations; is usually a scientist long absent from his discipline; often picks up the restaurant bill.
- *Ultimate Scientific Bureaucrat:* avoids discussion of business (science) at social gatherings, preferring to talk about golf, hobbies, etc.; more interested in your position and rank than in what you say; his assistant (gopher) picks up the restaurant bill.

To expand the key to enable rapid identification of the infra-groups within the *federal scientific career bureaucracy*, I have drawn up Table 1.

Obviously, with keys to subspecies and a table of characteristics, we are rapidly approaching a multidimensional matrix view of career bureaucrats—which is entirely proper. They, as individuals, are almost as varied as the general population, but in their work, their attitudes, and their world views, the variability decreases sharply and classification becomes feasible.

The Time-Server Bureaucrat

"Time-server" or "lifer" bureaucrats, as distinct from career bureaucrats, are only moderately committed to the job. They tend to be people who have learned to do assignments satisfactorily but not necessarily with excellence, and who have studied the system carefully to know precisely what their rights are in every job-related activity. They can be counted on to do a workman-like nine-to-five job at assigned tasks, but they cannot be counted on for enthusiasm, initiative, or personal involvement in issues.

At the lower end of a spectrum, the time-server bureaucrat enjoys protracted discussions during duty hours of sports or other non-job-related subjects, often has an overriding interest in a hobby, is an active member of the office rumor and gossip network, expects maximum recognition for minimum efforts, and has scant loyalties to the agency. Bureaucrats at this end of the spectrum tend to remain year after year in trivial positions, far removed from policy- and decision-making levels—doomed to a continued existence of background preparation, gathering information to be used by others for briefing and other kinds of documents, preparing draft responses to inquiries, and similar "scut work." For those who do somehow advance in authority, often because of sheer longevity, commitment to the job does not increase commen-

TABLE 1
LEVELS IN THE FEDERAL SCIENTIFIC CAREER BUREAUCRACY

Descriptor	Grade	Principal activities and characteristics
Novice bureaucrat	GS 11	Entry or gopher level; little command of exquisite bureaucratese; still somewhat idealistic about potential personal impact on "the system"; prepares background material for vast array of documents
Journeyman bureaucrat	GS 12	Much of entry-level optimism about flexibility of the system lost; may prepare drafts of memos or similar documents; may enter some office-level planning discussions if aggressive enough and bright enough; may attend some sub-policy-level meetings as an observer
Mid-level bureaucrat	GS 13	Has had marked success in self-hypnosis; depended on to produce useable drafts on many subjects; revises drafts developed by journeymen; asked to comment on drafts prepared by other mid-level people; participates in staff meetings, and may actually make comments
Emergent bureaucrat	GS 14–15	Keys to much of the technical and staff expertise reside at this level; finalizes documents for sign-off at higher levels; much time spent in discussion and planning
Senior bureaucrat	GS 16–17	Usually deputy or assistant to politician bureaucrat; a reservoir of expertise and whatever institutional memory exists in the agency; signs documents; provides advice to politician-bureaucrats in setting policy

surately. The "perks" increase and the freedom (in hours and travel and the choice of tasks to be accomplished) increases, but the bottom line is still to get through the day with the least possible effort.

COMMON CHARACTERISTICS OF BUREAUCRATS

Regardless of where in the previous three categories a bureaucrat may be placed, he or she will have at least a few basic identifying marks in common:

- Bureaucrats are most comfortable when they are with their own kind—those with compatible world views, sharing a common jargon, facing the same kinds of daily mini-crises, searching for relevance in a paper-dominated world, unable to shape or modify in any significant way the forces which control that world.
- Bureaucrats find solace in knowing that a system exists and is functional—that the system can, with good will, accomplish the job to be done.
- Bureaucrats know that minor perturbations in the system—such as periodic internally dictated reorganizations—are to be expected and sometimes even welcomed.

Beyond this core of identifying characteristics, there are special marks of *government* bureaucrats that should be fully understood by game players:

- They grasp all the nuances of interactions between politician bureaucrats and career bureaucrats, particularly in matters related to budgets and staffing.
- They recognize the critical nature of interactions with the Office of Management and Budget (OMB). They understand its relative autonomy and its totally prag-

matic viewpoint, within the confines of shifting executive policies.

- They are hypersensitive to all interactions with legislators, recognizing the fundamentally parochial interests
and the pet programs of such functionaries. They accord
particular attention to legislators who serve on budget,
finance, and appropriations committees, or on oversight
committees for their own agency.

- They are masters of the "three option ploy," in which
solutions to any problem occupy three standard niches:
(a) the do-nothing non-solution; (b) the moderate, effective solution; (c) the extreme solution, involving unpopular actions. Option (b) usually wins, nosing out option
(a) at the finish line.

- They fully recognize the sine-wave nature of government funding and spending attitudes—a repeated wave
pattern related to entry of new administrations, each
with initial vows to reduce government spending and
personnel. (The game player should attempt to avoid
vulnerability in grant proposals or renewals during
trough periods of the wave, since these are usually glum
times of retrenchment or uncertainty in the funding
agencies.)

- They rarely gamble the "whole bankroll" on any program, preferring to spread funds (and risks) over a number of areas, making everyone marginally satisfied but
no one happy.

- They understand that funding decisions are often political weapons, and they are sensitive to the need for analysis of all agency actions from a political perspective.

- They understand the delicacy of policy interpretation at
many levels; mild suggestions from higher administrative offices can be calls for prompt and decisive action at
field office levels.

- They fully understand that the degree of control of
events and destiny varies directly with ascent of the

bureaucratic pyramid. Near the bottom of the hierarchy inmates are abjectly at the mercy of the system and its vagaries. Even at mid-levels there are moments of total helplessness. Only near the top, at the political level, are there comforting feelings of being somewhat in charge, and these can be totally disrupted by changes in administration.

- They expect (but still dislike) major political perturbations, which can have drastic effects on the agency. Sudden budget reductions, termination of entire programs, restructuring of missions, major personnel changes—all can be fallout from shifts in political power.

The transition from one federal political administration to the next is an excellent time to view the rapidity and ease with which career bureaucrats respond to changes—even reversals—in agency policies. "Adaptation" and "survival" are key words in times of transition stress, with rapid assumption of new roles once the party line is established and marching orders given by new politician-bureaucrats. Good career bureaucrats charge off in new directions with enthusiasm, and within a matter of days or weeks the old directions and the old roles have disappeared from memory, never to be recalled, or to be recalled only in vague terms, in comparison to the bright new world of the present.

- They understand that one of the most important functions of a mid-level bureaucrat is to explain "the system" logically and plausibly to "outsiders"; and to support and protect the system, its policies, and decisions made by higher levels, regardless of personal opinions and preferences.

The large majority of mid-level career bureaucrats perform effectively and well, and are positive and enthusiastic about the organization in which they function. Achievement of a balance of reality in day-to-day operations and the idealistic principles which should and do govern attitudes and actions seems to represent a working objective of most bureaucrats.

THE BUREAUCRATIC HIERARCHY

The creation of interconnected boxes to categorize bureaucrats can be a meaningful doodling exercise in attempting to understand their roles and motivations. One of the most useful of such efforts groups bureaucrats according to principal functions—as producers, coordinators, or policymakers.

"Producers" are found at the lower hierarchical levels, as they are in any organization. They are the people who prepare documents, do background research, and interact with counterparts within the agency. An outstanding characteristic of incumbents here is acceptance of the deprivation of recognition and personal visibility for work accomplished. The compensation is a sense of narrow security, which leads to unwillingness to risk too much creativity or the dangers of movement to level two—the coordinator. Some bureaucrats spend entire careers at level one, until released by early retirement. Existing government systems do not provide adequate rewards of "rank without risk," so security becomes more important than creativity at this level.

"Coordinators" or *"organizers"* at level two are those bureaucrats who have accepted the risks inherent in greater visibility and decision-making. They are the heads of offices and various subdivisions of an agency. They are often very effective in their roles, and they are usually cautious gamblers, knowing that once the security of the producer box is left behind, there is no return—the path leads up or out. They know too that successful interaction with level three—policy level—people is crucial to survival.

The final category, the *"Policymaker"* or *"Limelight"* level, is characterized by temporary power, large plush offices, and the

stimulation of close contacts with other exceptional government people—elected or appointed—as well as with public and
industry groups. Bureaucrats at this level are all gamblers; they
fully appreciate the transience of their reign, but most of them
believe that they are good enough to move to roughly comparable future positions within or outside government. Occu-
pants of this final box may be effective executives, but they must
be politically based, which almost guarantees a short life span
in any particular position.

A flow diagram of bureaucracy, from creation to oblivion,
using the perspective of the preceding paragraphs, might look
like Figure 3. The diagram indicates that opportunities to enter
or leave the system from and to the outside exist at each level,
but also that opportunities for reciprocal lateral movements
between boxes do not usually exist. Security-oriented people
stay in the level-one box; gamblers move from level one to
level two, and (very rarely) even to level three, if they are
very good and have adequate political/industrial support.
Hovering over each level is the threat (or the promise) of the
"outside"—an amorphous zone which can include private industry, academia, other agencies, early retirement, or the
unemployment office.

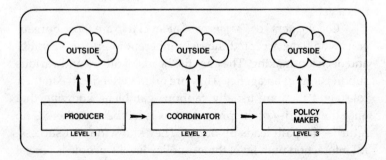

FIGURE 3. A much-simplified summary of maneuvers available to the
bureaucrat.

KINDS OF BUREAUCRACIES

Bureaucracies come in almost endless varieties, so any attempt to categorize them can be frustrating. It seems important, however, for the scientific game player to recognize the principal varieties, especially those which impinge directly on the conduct of science. I have elected to discuss here three types of bureaucracies that are of direct concern to scientists. (1) government science management bureaucracies, (2) scientific administrative bureaucracies—university and industrial, and (3) bureaucracies created to disburse research funds—public and private. There are, of course, other bureaucracies which impinge on everyday existence—regulatory bureaucracies, service bureaucracies, and administrative bureaucracies, in particular—but these are not integral to the objectives of this book.

Government Science Management Bureaucracies

There are many major in-house research and development programs in government. These are mission oriented and usually long term. Examples include agricultural research, communications systems research, forest research, health-related research, fisheries research, and a host of others. While significant parts of the total government research effort are actually conducted by private contractors or by universities, the in-house research is substantial and extensive. Management of this research is dispersed among many subdivisions of the federal executive departments, and is entrusted to an uneasy but viable mix of politician-bureaucrats and career bureaucrats. The managers at the agency level may be former scientists, or they may be drawn from a variety of professions such as law, business administration, or economics. Field-level managers of research units are almost always former scientists, often with appreciable stature in their disciplines.

Individual professionals encounter government science management bureaucracies when they join a government agency, when they accept a short-term assignment (usually as part of a sabbatical with a government agency), or when they accept a research grant or contract from a government agency. Initial reactions to such encounters are varied in the extreme. Some scientists are appalled by the energy and funding devoted to managing, planning, and evaluating, rather than *doing* science. Some are totally frustrated by the sheer volume of paperwork required. Some are delighted with the positive reaction of government managers to expressed enthusiasms and infusions of fresh expertise. Some are impressed with the variable but often excellent quality of the in-house scientific staff.

The game player usually finds brief encounters with government science management bureaucrats and bureaucracies to be intense learning experiences. The variety of world views and perceptions of scientific contributions to society can be mind-stretching.

Scientific Administrative Bureaucracies— University and Industrial

Whenever there is a critical mass of scientists and scientific research (interpreted as more than two working professionals), some form of management structure will be superimposed. Whether the scientists are members of a university faculty or of an industrial research team, someone will have perceived the pressing need for proper management of the group. The beginning is simple—a chairperson or group leader or dean, then a secretary, then an assistant leader and an administrative officer, then more secretaries and a typing pool, then personnel, accounting, and purchasing clerks . . . on and on. Eventually the group must be subdivided into divisions, each with a chief and deputy, and each with appropriate administrative support. If the organization functions in a matrix mode, task leaders and assistants must be designated and proper administrative support assigned.

A worrisome aspect of administrative bureaucracies of this kind is that they gradually assume *control* over those (the scientists) whom they were created to *assist*. Administrative groups of this kind become ends and independent self-serving entities in themselves, and may delude themselves into thinking that the scientists are expendable, or at least only a minor component of the organization.

So in addition to coping with external, usually governmental, bureaucracy, scientists must accommodate to or resist their own organizational administrative bureaucracy—whether it be university or industrial. This bureaucracy consists of all the administrative structure that is created ostensibly to ensure that people work, that research is carried out, and that a product is developed. It is usually populated by nonscientists who have a persistent focus on dollars—and this is the root of most of the problems, since scientists as a group do not share or want to share such an outlook on science.

Bureaucracies Created to Disburse Research Funds

As government has moved more and more into the funding of science during the past half-century, scientists have had to learn first about the existence of, then how to deal with (and to live with), bureaucracy and bureaucrats. Bureaucrats are part of a complex system; scientists understand complex systems in science; therefore scientists should be able to understand and manipulate the bureaucratic system to serve their own purposes.

Of course there are fund-disbursing bureaucracies other than governmental, but federal funds will probably continue to be a major source of research support, so most of our discussion will concentrate on that source. There are few simple approaches to the myriad ways in which bureaucracy interacts with science; the best I can do here is to take examples of such interactions, and to suggest some tentative rules of the game that must be played.

One basic overriding dictum that governs any interaction with funding bureaucracies and their resident bureaucrats is the necessity for great gentleness and patience. Overt indications of pity or condescension for the role in which such bureaucrats find themselves should be avoided. Many scientists-turned-bureaucrats or administrators still fancy themselves as scientists. Do nothing to splinter that image; instead, cater to it, avoiding references or questions which might display their ignorance or scientific obsolescence.

Usually, old-line bureaucrats in funding agencies do not like to deal directly on a one-to-one basis with scientists. They are much more comfortable with a shield of consultants, advisory committees, and review committees deployed out ahead of them. Occasionally, relatively junior agency staff members with backgrounds in science will participate in face-to-face encounters with scientists, but not the more senior agency staff (with some notable exceptions). Since such minor bureaucrats can have great influence with their superiors, people at that level must be treated with reasonable respect.

The complex network of federal grants and contracts has called into being an entirely new specialized area of study called *grantsmanship*—the art of acquiring federal (as well as corporate or foundation) funds. Grantsmanship is an important aspect of game playing—a critical survival factor in today's highly competitive funding world. A few key suggestions for introductory grantsmen or grantswomen are these:

- Learn the proposal jargon and format of government agencies and private foundations.
- Know the interests of the funding agencies and slant the writing of your proposals to such interests.
- Have your proposal scrutinized by a professional proposal writer if one is available (most large universities have at least one such person on the staff).

- Hire yourself out to an agency which makes grants—government or private—for a short term, even on a volunteer basis. The time might be well spent, and you might even find a whole new career there as a science administrator.
- Take time to think carefully about who might be logical reviewers of your proposal, and be sure that their work is recognized and praised in the "background" section of the proposal.
- Remember that proposal preparation is an art, but most of its elements may be learned. While there is no real substitute for ideas, insights, competence, and accomplishments in research, these elements may not always come shining through to reviewers of your proposal without substantial effort in its preparation.
- Always ask for one and one-half times the funds that you actually expect to receive.

Bureaucracy requires the existence of someone or some group to whom the buck can be passed if the occasion demands it. If such an entity does not exist, then it will be created, to act as a heat shield. This is nowhere more apparent than in the grant-approval procedures developed by many federal agencies. Such procedures include often elaborate *review processes* carried out by elements of the scientific agencies. The basic concept—evaluation by peers—is good, even though it is surrounded by abuses (which I will get to in due time). Most scientists would rather be judged by peers than by clerks, politicians, or accountants.

Since the review procedure looms so large in grant approval, and since these procedures seem to form a workable if flawed shield for bureaucracy, they are apt to be permanent fixtures, hence must be understood and accepted as an element of life in science.

One review procedure that has proven effective is the so-called site visit, occasionally by a staff member or team from the granting agency, but much more effectively by a committee of scientists generally knowledgeable in the subject matter area of the grant proposal. Often these committee visitations are to consider large, multidisciplinary grants of sizable sums, so the university hierarchy must be interested and involved. The site-visitation committee provides great opportunity for personal interaction, and naturally lends itself to consummate game playing—so much so that a set of strategies should be developed by the grant proposer. Here are some suggested strategies:

- If several faculty members are involved, and particularly if they are from different departments, have a dry run two weeks in advance of the visitation to check the firmness of presentations and adequacy of visual aids, and to eliminate the uncertain, the boring, and the excessively nervous individuals. Be sure that the cohesiveness of the proposal emerges, and that all investigators are familiar with the proposed research of the others.
- Conduct a preliminary briefing for junior faculty members involved in the proposal, in an attempt to avoid unnecessary bruises and feelings of rejection. Point out that major pitfalls are attempts at "snowjobs," lack of enthusiasm, dogmatic authoritarianism, and failure to do homework.
- Spotlight those who are knowledgeable and make a good appearance and presentation, and hide those who do not display these characteristics.
- Remember the necessity for directly identifying and addressing the chairperson and the true decision-making members of the visiting group—and ignoring the occasionally vocal but noninfluential members.
- Respond to questions—often abruptly interjected—in a reasonable, calm, and *brief* manner, without undue interruption of the presentation.

- Cater to members of the visiting committee—all of whom have scientific or administrative competence, many of whom may be authorities in particular subdisciplines, some of whom may want to make statements indicating their competence and authority to other members of the committee (whom they may not have met previously).
- Remember that the visitation may have elements of stress for the visiting group—they are expected to make rational recommendations, act informed, avoid disclosure of personal areas of ignorance, and stay awake.
- Allow more than adequate time in the agenda for questions, mini-lectures, and pontification by review panel members.
- Have as high an organizational official as possible make the introductory remarks, *provided* that he or she is knowledgeable about the proposal.
- Prepare and distribute in advance to all university participants a fact sheet on each review panel member, including areas of competence and major publications.
- Be sure that coffee and pastries are available on arrival, and that coffee is available continuously.
- Use two-sided place cards with large lettering for each review panel member.
- Do not underestimate or underplan the activities at the inevitable and highly essential evening cocktail hour and dinner. This is the time to bring in your advisory board members and other community supporters, if any.

The site visit can be a pleasant, relaxed, informative period, or it can be a tense and uncomfortable period, depending on the attitude of the principal investigator (or proposal coordinator) and on the extent of the advance preparation by the proposal group.

I promised earlier to list some of the evils of the review process. They include the following:

- Sometimes, for political or other reasons, the good and conscientious efforts of peer review boards are virtually or absolutely ignored by the funding agency in making final decisions about grants.
- Many funding agencies have a closed circle of science administrators and scientific consultants who shape the nature of the review process and hence the decisions about grants. This "good old boy" group is self-perpetuating in that when vacancies occur or terms expire, replacements are drawn from friends or former students.
- Grants often seem to flow most easily and most often to former recipients of grants, and rarely to relative newcomers. This may be a logical progression, since investments are safer with proven performers than with novices, and since there is a learning curve in preparing grant applications. It is, however, a real gripe among junior faculty members.

After the battle for funding has been won, there is still the problem of guerrilla warfare within the grantee's own organization. Hard-won grant dollars can be manipulated by university bureaucrats—even at the most prestigious schools. Sometimes the way the dollars are proposed to be spent in the grant proposal and the way they are actually spent by administrators are quite different. Reported variations include salaries for people who have nothing to do with the grant, and siphoning off more than the amount proposed for university overhead. Advice to game players is to watch for variations from proposed spending, even though such things as budgets are something of a bore. Have an accountant friend look over copies of the university financial statements or reports of expenditures on your grant. If you don't regularly get such reports, insist on them.

Be just a little diplomatic about all this, though, lest you become known as a troublemaker, or worse still, as a "whistle blower." Organization bureaucrats have set prescribed modes of action and are easily upset when their routine is disturbed by absurd requests (like an accounting of how your grant money is spent). Be reasonable but firm, and if too much resistance is encountered, see the dean or chancellor for advice and possible assistance.

SUMMARY

If bureaucrats did not exist they would have to be invented, since they provide continuity and stability in any agency or organization. Bureaucrats can be identified by the fact that they deal with paper and words rather than with products. Interaction with good bureaucrats on their home turf is one of the most challenging activities that a scientific game player can engage in, so preparation for the encounter is important. Recognition of the types of bureaucrats and the kinds of bureaucracies is part of that preparation—as is some understanding of forces within the bureaucratic universe. Good bureaucrats are survivors of a severe selection process; they often find satisfactions in their jobs that are equivalent to those found by scientists; and most importantly, those at decision-making and policy levels can have significant influence on science. Do not underestimate them!

CHAPTER 12

DEALING WITH EXTERNAL FORCES
News Media, Lawyers, Politicians, and the Public

The news media and their problems; lawyers as climax predators; politicians as necessary evils; the man in the street and his perceptions of science.

If you were to ask the average television-watching, newspaper-reading, vocal middle American to describe his concept of a scientist you might get back some of the following: "truthful, cautious, obscure, occasionally pedantic, sometimes verbose, unable to say yes or no without qualifications, and unwilling to take firm stands." In recent years, though, more and more scientists seem willing to stand up when counts are made, and to use public forums for clear statements on popular and unpopular issues. This trend is laudable and proper, but public interaction can be a dangerous arena for the amateur. Dealing with the public and its spokespersons—news media, lawyers, and politicians—successfully is an occupation for professionals, and one calling for superb game playing. As such it is well worth consideration in this book.

NEWS MEDIA

Scientists who have something to say to the public obviously must sooner or later interact with representatives of news media—principally newspaper, television, and radio reporters. This interaction can lead to effective public release of sound scientific information or to garbled misrepresentation of facts and information. Usually the outcome is determined principally by the skill of the scientist, and to a lesser degree by the skill of the reporter.

The trauma of negative results in attempted communication with representatives of news media can be lessened or eliminated altogether by some small attention to a set of guidelines which have been developed painfully during successful and unsuccessful encounters with reporters and feature writers:

• Always keep in mind that the principal job of a reporter is to develop a story that will sell newspapers or attract and retain television viewers or radio listeners. To this end the reporter will not usually deliberately misrepresent what you have said, but may emphasize what to you are minor points, may weave your comments in with those of others, may use part of your statement out of context, or may eliminate many or all of the qualifications that you have placed on your conclusions. Then the headline writer may create an erroneous impression about the already distorted content, leading you to complete frustration and to a pledge never again to give even the correct time to a reporter.

Much of this can be avoided by careful attention to your statements, by avoiding technical language, by repeating your key points, and by absolutely refusing to speculate or to go beyond the safe confines of your data in drawing conclusions.

• One of the best approaches to at least satisfactory interactions with reporters is to select a few who are basically interested in science and to offer them five-minute doses of intensive science education in relevant subject matter areas. Admittedly they are pressed for time like everyone else, but they like to feel occasionally that they are on top of a story and really understand what they are writing about. After a few such sessions, the likelihood of a story's appearing that will have you climbing the walls or mumbling to yourself will decrease.

Larger newspapers often have good science reporters, as do public and some commercial television stations. I have dealt over the years with several such informed, perceptive, and highly intelligent people, and the results have usually been good. Problems develop when uninformed (and often uninterested) reporters are assigned to a scientific story as just another job, and handle it as just another job, without insight or enthusiasm.

Some of my best experiences with news media representatives have been with very good science feature writers with a large, southeastern newspaper. Detailed research, crisp, accurate writing, and reasoned interpretations of science affairs by these writers made continued interaction with them over a period of several years a pleasurable and mutually satisfactory association. I knew that they weren't going to distort the facts given them or draw conclusions that weren't warranted by the data. I knew too that I could talk to them openly without the danger of being misquoted or being drawn against my wishes into controversies with other scientists.

This pleasant state of affairs is rare. Scientists are well advised to choose words carefully when talking to media people, since, at their discretion, "anything you say may be used against you" or cast in a context other than that in which the words were used.

• Scientists are often approached by feature writers to supply material for a Sunday supplement or other article or series. Such writers obviously play on the vanity that exists in most of us—to see our names, our research, or our results in print or on television. While such feature stories can be an important way of getting scientific information to the public, the scientist often has no control of the content, composition, or tone of the story, even though the principal contributor is the scientist. As a group, scientists are too reluctant to set down any guidelines or stipulations to feature writers—such as the very important right to review and comment on the article before publication. Usually the reporter pleads lack of time before deadlines or violation of tradition, but in reality these stipulations can be made if we are willing to risk a little vanity by having the reporter walk away—a small loss, considering other available public outlets.

An almost classic example of abuse and distortion of information given freely by scientists to feature writers concerned a Sunday magazine article on aquaculture. A writer and cameraman visited a small pilot-scale aquaculture project, asked appropriate questions, and received published (and some unpublished) information on the project. They then visited several other similar projects in various parts of the country. What an excellent opportunity for an enlightened, realistic article on the present status of aquaculture—and what a hash was made of it! The article appeared a few months later with a headline suggesting that aquaculture was a panacea for world hunger. The article itself consisted of extrapolation after extrapolation of minuscule pilot-plant production figures and unproven cost estimates to future large-scale successful commercial ventures—the very last kind of interpretation that should have been made of the data. A letter to the managing editor went unanswered—but, no matter, the damage had been done and the reading public had been misinformed once more.

• If a news story appears that garbles, distorts, or misrepresents scientific information which you have provided, there is every reason for communication with the managing editor, stating your opinion calmly. The public needs good, reliable reporting on scientific affairs; neither they nor we need to tolerate hacks.

Be careful, though, not to be drawn into a public controversy with other scientists by news reporters unless you are ready to stand the heat successfully. News media thrive on controversy, and they react with great glee when a piece of bait is swallowed—especially by experts. Take pen in hand only after suitable introspection; use the phone to express your outrage. This dictum is particularly relevant if you are expressing the opinion of a scientific group or government research agency.

LAWYERS

One of the favorable outcomes of the environmental movement has been the emergence of independent, courtroom- and hearing room–wise scientists who have the background and facts to hold their own with industry scientific consultants, with administrators or consultants from enforcement agencies, or with government scientists. Adversary proceedings, though they may appear unseemly to some scientists, do have the salutary effect of ensuring public exposure of *facts,* as well as conclusions drawn from these facts by various experts.

The game player, if he or she is to enter the hearing room or the courtroom, needs several high-level qualifications:

- An established reputation as an expert, with a clear and current expertise in the area of controversy, and a complete familiarity with published information;
- More than passing acquaintance with legal concepts and courtroom procedures—particularly with hearsay information, data and its abuses, and limits of expertise; and
- Painfully acquired *experience* in dealing with lawyers and with their legal universe and world views, which are largely foreign to most scientists.

Some useful approaches for the scientific expert include auditing selected courses in law schools and attending relevant hearings and cases as an unidentified observer. One of the most difficult of legal concepts for the scientist to accept is that *lawyers want to win cases,* and are not concerned unduly with a search for truth unless it is of advantage to their case. It can be a great source of frustration to uninitiated scientists to find that their good and conclusive information may never be allowed to be presented, or that their testimony is brought into question by seemingly extraneous considerations, or that they may be discredited as experts by technicalities. Fortunately, most scientists live and die without having to appear as expert witnesses; for those who do become involved, the ultimate in skill will be required. Basic procedural steps include the following:

- Establish *competence,* as indicated by training, degrees, experience, research, and publications.
- Establish *credibility* by demonstrating that data and conclusions are based on the use of standard methods, producing results within acceptable confidence limits.
- State *facts and conclusions* drawn therefrom clearly, distinctly, and succinctly.
- Offer clearly identified *opinions and interpretations* if requested, but never extend beyond the confines of the best available information, regardless of pressures.
- Where possible, *insert a carefully prepared, scientifically sound, signed document* containing a summary of your remarks into the record.

The prudent scientist would be well advised to treat the courtroom or hearing room as a jungle, peopled by such climax predators as (1) lawyers who would impugn or destroy scientific reputations without a moment's hesitation, (2) hostile expert witnesses (the "scientific streetwalkers" mentioned in the chapter on ethics in science) who have the credentials and the savvy to cast doubt on the testimony of the uninitiated, and

(3) hearing examiners who may have an agency position to defend, regardless of the evidence.

Encounters with legal minds range from informal hearings on patent applications through formal hearings conducted by agency examiners or fact-finding hearings conducted by district courts, to subpoenaed testimony in civil court cases involving claims for damages from such events as toxic chemical spills or vessel groundings. Some ground rules apply to all: never underestimate the intelligence of any person involved; never make comments for the record which cannot be supported by data or which are beyond your area of expertise; don't give the court a mini-lecture in your area of expertise, regardless of the temptation; where possible, come with a prepared signed statement, to ensure that your complete testimony will become part of the record in the event that you are prevented from giving it orally; make certain that every word of the prepared statement is defensible and founded on fact; and if possible have the prepared statement scrutinized by a lawyer before it is submitted.

For those scientists with genuine expertise, with dedication to public disclosure of facts, with willingness to learn, with agile minds, and with a suitably thick and keratinized epidermis, here is a stimulating and demanding testing ground.

POLITICIANS

Scientific game players and politicians have much in common and should interact easily. Politicians want to be re-elected, preferably in a higher capacity; scientific game players want to progress rapidly in their chosen field. Politicians would like to make some lasting impact in a few selected areas; scientific game players have similar objectives. Politicians recognize that people interactions are the key to success, as do scientific game players.

Despite these and other similarities in philosophy and objectives, scientists and politicians share a high degree of mutual distrust. Politicians usually feel that scientists are "ivory tower," given to impractical solutions, and totally unreliable as "team players." Scientists view politicians as ultimate pragmatists, users and abusers of people and expertise, and in a biological category given to coprophagy.

Faced with significant commonalities but mutual disrespect, how do politicians and scientists interact? (Answer: "Gingerly, if at all.") There are areas of interaction that can be exploited by the scientific game player, if he or she is in the proper position:

- Politicians occasionally need scientific information or advice that they can count on as being unbiased and definitive. Contacts with the proper legislative assistants can create a niche for the informed scientist interested in the political process.
- Politicians are always looking for areas of special interest or concern in which they can become leading proponents or opponents. Some of the scientific areas (energy, pollution, population, environment, endangered species) with high public visibility are in this category and can be subjects for dialogue between scientist and politician.
- Scientists need proponents for large-scale research projects—proponents who can help obtain funding for such endeavors. Politicians can fit such roles, if they are convinced of their worth or of the advantages that will accrue to themselves by supporting such projects.
- Scientists need support from politicians in debates on public issues. Public expression of concern and presentation of scientific evidence are of little value without the perception of an issue by politicians and their willingness to translate expressed concerns into legislative action. The scientist must be prepared, however, for the withdrawal of political support for any issue, based on factors that may be totally unrelated to the scientific significance of the issue.

The scientific game player will develop and maintain some form of communication, and even rapport, with politicians, especially at the state and national level. National legislators usually have at least one staff assistant particularly responsible for scientific-technological matters, and this is the person to be cultivated, in addition to the legislator. Such contacts can lead to requests for background help in drafting legislation and speeches and can often be of mutual benefit in unanticipated ways (in favorable response to a request that the legislator be a society banquet speaker, for example). The political process is enormously complex, and legislators face a bewildering array of advice and demands on countless issues. The opportunity to provide reasoned scientific input is one to be cultivated with vigor and intelligence.

THE PUBLIC

The same unreliable polls that resulted in the description of a scientist given earlier in this chapter also disclosed that people in general want to like and trust scientists but are confused by obscure and often conflicting statements attributed to scientists as a class. Furthermore, they do not want decisions about their lives, or the lives of their children, to be completely in the hands of scientists (or politicians, either). They want to understand enough about technology to participate knowledgeably in the decision-making process.

The scientific game player, of course, has long known about these reservations and concerns, and acts sensibly to become a reasonable interface (another lovely bureaucratic word) between science and the public. The avenues for success in this interface zone are numerous and tempting for the consummate game player:

• Actions by industries and inactions by government that result in further degradation of the human environment rank high in most polls of principal concerns of citizens today. The confusion level is high because of often-conflicting conclusions and interpretations of data by scientists and regulatory agencies. There is a genuine need for reasoned, unbiased evaluations by good scientists, either individually or in groups. Unfortunately such evaluations must be paid for, and people are sensibly suspicious of any findings that seem to favor the donor of funds.

There is no simple solution, but one chosen by many game players is to be on the side of the angels. A number of scientists have become effective spokespersons for environmental concerns without unduly prostituting science. In this role they have developed entire new careers as evaluators of data produced by others, as producers of data in critical areas, and as writers of popular and semitechnical books. These scientists have become in some ways equivalent to the pamphlet writers of the Revolutionary period in United States history. Today's "scientific pamphleteers" attempt to arouse individuals and the country to action, just as their revolutionary predecessors did.

There is always a clear danger, however, of completely or partially losing objectivity in adopting a protagonist role. The same danger exists in assuming an antagonistic role in environmental matters. The boundaries of the playing field become obscure where conclusions and recommendations may be drawn from the available data, but where *interpretations* can be very divergent.

• A number of scientists have emerged from the large general science lecture halls and the less-than-prime-time early-morning educational television series to become television personalities, speaking on matters of concern to the public in commercial television appearances. This is a logical and significant progression for the few with the high qualifications of appearance, knowledge, and ability that fit them for the role of "scientific interpreters" or "scientific translators." Unfortunately, some of these chosen few tend to become overly convinced of their own infallibility and begin sounding too much like prophets and too little like scientists. If such a natural human tendency can be resisted, and if the scientific soapbox of television can be handled rationally and effectively, much can be done to educate and guide the average person.

• Scientists often find themselves in roles of advisors to politicians, from the President down to the small-town mayor, and to innumerable committees, councils, and commissions. This role can be an effective one, provided that the advice given is really listened to and that the scientist is not merely window-dressing or a facade behind which preconceived actions are taken.

This of course does not imply that scientific advice will always be the ultimate determinant of a decision. In fact it rarely is, and many scientists are appalled to find that their advice constitutes only one of a large array of factors—economic, social, and political—that are considered in reaching decisions. Genuine and enthusiastic involvement in such decision-making processes by "scientific consultants" or "scientific advisors" is still preferable to noninvolvement.

SUMMARY

Scientific research, as an approach to development of understanding of the universe, depends on the financial support of nonscientists for its continued existence. Because of this, scientists must interact occasionally with representatives of the outside forces which help to shape the selective process of survival. These representatives—principally news people, politicians, and lawyers—are bright, perceptive, and aggressive—and have objectives of their own. They may use scientific information or expertise if it is to their advantage to do so, but generally they find most scientists to be obscure and somewhat unworldly. It is not surprising therefore that the few scientists

who make sincere attempts to relate to laypeople or to specialists in other professions are cultivated and listened to. It is not
surprising either that such attempts sometimes result in frustration when the message is poorly transmitted or improperly
received. Some communication is necessary, but special talents
and behaviors are required of scientists who are successful.

CHAPTER 13

THE SCIENTIST IN INDUSTRY

Some pitfalls in industry scientific research; industrial game rules; kinds of industrial research laboratories; consulting; the scientist as an industry spokesperson.

The three principal customers for scientific expertise are universities, industries, and governments. Throughout this book we have considered ways in which the perceptive scientist finds his or her way through government bureaucratic and university/academic systems, but we have paid too little attention to the most demanding, most ruthless, and most financially rewarding employer of scientists—private industry. This chapter will attempt to provide a relief map of that terrain, as viewed from inside and outside an industrial research group.

It must be pointed out at the outset that industrial research is different from other kinds of scientific research in a number of ways:

- It is almost totally product–oriented except in the few largest organizations that can afford some basic science.

- It is much encumbered with documentation of all kinds—planning documents, progress reports, quarterly reports, annual reports, etc.
- It is much more conscious of and responsive to benefit-cost ratios and other forms of financial evaluation and controls.
- It is more heavily committed to team research, within a matrix type of organization.
- Its salaries are larger than those for comparable positions in universities or government.
- Its projects are much more subject to rapid and at times almost capricious termination, with the associated risk of sudden unemployment.
- Some of it is farmed out under contracts to academic consultants who are leaders in their specialties. The contracts always specify confidentiality of findings and usually call for company ownership of all patent rights.
- Industrial research scientists are almost invariably shackled by contracts prohibiting publication in the open literature without corporate approval.

Despite the disadvantages of industrial research, the pay scale and the availability of good and often unique equipment continue to entice scientists in great numbers. Once inside, though, the scientist faces numerous difficulties. The principal one is that engineers and accountants are usually in control or otherwise dominate the decision-making process. Such people are concerned with systems, schedules, and profits, and with justifying research on a dollar-for-dollar basis, which is almost impossible. Another difficulty is that some industrial research groups will deliberately overstock neophyte scientists, then carry out large-scale dismissals of those who for whatever reason do not seem to be making it. This kind of ruthless tactic has long been practiced with engineers and certain other professional categories; it results in a tense "up or out" atmosphere that many young scientists cannot or will not tolerate.

INDUSTRIAL GAME RULES

Despite all its drawbacks, industrial research can be attractive to bright, aggressive, well-trained junior scientists—especially those who are sensitive to game rules as they apply to this research environment. Among many such rules, these seem of significance:

- Establish an early reputation as a good team player. Any matrix type of organization depends for success on the goodwill and initiative of team members rather than on line authority.
- Charge ahead, but give others appropriate credit.
- Be careful to avoid the "Rumpelstiltskin" or "Gopher" syndrome referred to in Chapter 7. This does not mean avoidance of hard work or long hours, but it does mean receiving adequate credit for the work you do.
- Be particularly mindful of the patent possibilities for processes or equipment that you develop. If company policy permits, begin patent procedures on anything that seems even remotely patentable. This helps identify you as a "comer" and gives you some visibility in the parent company. Often, though, the employment contract will specify very clearly that patents become company property.
- If a particular project seems to be going nowhere, or if it is headed by an unaggressive team leader, take the first opportunity to move to another project. Do not do this too often, though, since repeated changes may suggest lack of interest or inability to carry through on an assignment.
- If you have science management as a career goal, begin early to acquire training, particularly if the company has a plan—and many do—for released time and tuition payment.
- Any unauthorized release of information, or any statement that could be construed as "whistle-blowing"

should be approached with great caution, considering the ugly consequences and career derailment that could result. Corporations have long memories and deep pockets.

Beyond these generalizations about the ways in which industrial research differs from other kinds of scientific research, and about some hardball industrial game rules, it could be instructive to look next at some of the usual types of industry research groups that a scientist might join.

KINDS OF INDUSTRIAL RESEARCH LABORATORIES

I have selected three of the most common types of industrial research groups for treatment here: the *large industrial research laboratory*, the *new acquisition of a diversified corporation*, and the *small entrepreneurial venture*. (The special case of consultancies, which, like the camel, are partly inside and partly outside the commercial tent, will be given separate status later in this chapter).

The Large Industrial Research Laboratory

Large industrial research centers have many favorable attributes. They are often located in rural campus-like settings, headed by directors and team leaders with genuine scientific stature and credibility, stuffed with the latest and most complex equipment, and liberal with technicians, advanced training opportunities, and meeting travel funds. Why, then, do so many scientists dislike or hate them? Some of the reasons given are that such research groups are often controlled by financial managers, regardless of the titular head; that they can be profoundly capricious, phasing out projects or programs with minimum provocation, regardless of merit or productivity; and that they are users and abusers of professional people at all career stages, but particularly at the introductory levels. The campus-like environment can mask intense power struggles, vicious competition among research groups, and blatant manipulations of research data to serve the company's goals. Research groups can be used as fronts or covers for destructive environmental activities by the parent company.

The New Acquisition of a Diversified Corporation

Scientists are sometimes recruited to expand the technical base or competence of a new production component of a large diversified corporation. Examples of this are abundant. A large, metal-manufacturing concern decides to begin an aquatic-farming operation; a large animal feed manufacturer decides to invest in a vaccine-production venture; or a large soft-drink company decides to form a desert plant-production farm. All of these require a high level of staff technical competence, particularly in the early developmental years. This means that salaries are usually high, in order to buy the critical talent. Scientists are hired, facilities are built, and equipment is bought— all without serious thought of profits, since the parent company can use the new venture as a tax write-off for several years.

Eventually, though, accountants and budget analysts prevail and the fairyland begins to decay. Words such as "benefit-cost ratios," "scale up to production levels," and "profitability" begin to be heard. The honeymoon begins to go sour, rapidly or slowly, with the departure of some key technical staff members; or a corporate decision is made to terminate the venture almost overnight, regardless of the extent of technical progress being made; or a corporate decision is made to sell the venture, just when it seems on the edge of profitability.

Fortunately, technical ability and experience are transferable commodities, but the scientists involved in such corporate manipulations have to be a special, hardy breed to survive drought periods. The usual trend, even for the hardiest scientist, is to retreat sooner or later to the comparative stability of universities or government laboratories. To the credit of such scientists, however, is the slow increase in the required technical base for the kind of product being developed—eventually to the point where processes become standardized or mechanized and profits can be made by successors.

Diversification seems to have become an operating principle of some large corporations, particularly during the past several decades. Sometimes diversification results in acquisition of small, technically based companies by a corporate giant—possibly because of the prospect, based on economic analysis, of large future profits, or because of a whim of one of the stockholders or officers. The former process recently led a large chemical company to acquire a small, high-technology fish-producing operation in the Southeast. Scientists and technical staff members were hired at excellent salaries, facilities were expanded, and optimistic reports (in color on slick paper) described the future of the operation and the production processes being developed. Then, during the third year of operation, and still several years away from a break-even point, a prolonged cold wave destroyed most of the animals. During the following spring pesticides were detected in the intake water, and the federal EPA imposed severe restrictions on the quality of the effluent. This was too much for the distant corporate officers and stockholders. The operation was closed, despite major investment in facilities and equipment; the technical staff was given two months' severance pay; and the facility stands today abandoned and rusting in the Florida sun.

The Small Entrepreneurial Venture

A time can occur in the lives of many scientists when they may become bored with routines of teaching or grant-proposal writing, or dissatisfied with their rate of advancement where they are, so they may be tempted to join or to help form a new venture based on applied science. There are numerous examples, particularly in the neighborhoods of large universities, of successful, rapidly expanding companies that began this way. There are many more examples of small, technically based ventures which did not survive, usually because of inadequate financing, poor management, inadequate markets for the new product, or inadequate technical development of the product.

Scientists considering participation in such ventures usually recognize some of the risks involved but are often blinded by the potential income advantages if the company succeeds (and some do). A balance sheet should be prepared that would assess the likelihood of success within a reasonable range of error. Components of the balance sheet should include (but are not necessarily restricted to) the items shown in Table 2.

Even with such a carefully completed balance sheet to aid in job decisions, there is always the possibility that unexpected

TABLE 2
A Survival Checklist for Scientists Contemplating Joining
a Small Business Venture

	Adequate	Marginal	Inadequate
Availability of working capital sufficient to survive several years without profits			
Realistic estimates of the market potential of the product			
Status of the patents on which the product depends			
Availability and quality of the managerial/financial talent within the company			
Likelihood of rapid development of competitive products by larger companies			
Degree of independence from public funding (EDA grants, federal contracts)			
Status of technical development of processes on which manufacture of the product depends			

or unknown factors will affect survival of the venture. New government regulations, changes in prices of critical materials, embezzlement, legal decisions about patent rights—all can destroy or seriously cripple an otherwise viable operation. Often, too, economic analyses based on faulty technical assumptions can be proven incorrect by bankruptcy.

One case history, which I am sure could be repeated a thousand times with different circumstances and actors, concerns a small aquatic farming venture raising shrimp in the Southeast. Capital was marginal but was augmented by federal (EDA) funds to permit lead time for technical development. Scientists (biologists) were hired, and engineering talent was available on contract. During the first year drought reduced the carrying capacity of the ponds, and unfavorable salinity reduced survival of the young shrimp. The second year's crop was growing well until a disease outbreak destroyed over 90 percent of the animals. The third year's crop was modest and did not approach profitability. At this point the original entrepreneurs quit, giving the staff a month's severance pay and selling out to an importer of tropical fish.

A common event in the developmental history of a small entrepreneurial venture is its acquisition—along with its patents—by a larger company with production emphasis in a similar application area. In pharmaceuticals, for example, the larger company acquires the patents and the professional staff of the venture company and can move, because of better funding, more rapidly through testing and approval of new drugs. An alternative to this course of action would be the hiring away by the larger company of key creative scientists from the venture company. Those scientists may take critical processing information with them, or copies of experimental formulations, or in some instances even experimental materials. They then continue development of new technology for their new leaders in a company with greater financial resources.

CONSULTING

Industrial research is often performed by outside consulting firms. The arrangement may be by contract for a single study or it may be a long-term association perpetuated by a series of contracts. Whatever the arrangement, there are two management levels to be satisfied: that of the contracting company and that of the consulting company. Scientists report to the appropriate supervisory level of the consulting company, and their work is incorporated into the final contract report. There is little opportunity for or encouragement of individual scientific publication.

Consulting is a high-risk occupation, and the risk exists for all employees of consulting firms. The flow of new contracts must be continuous to support the existing staff; any reduction in contract flow usually results in layoffs.

It is difficult to see why any sensible junior scientist would join a consulting firm except as a *partner*—in which case he or she would share in the profits as well as the risks. Otherwise, employment with such firms might be satisfactory as a stop-gap, short-term measure, until something more suitable developed.

Those who form or head consulting groups, however, are a different breed. They are the entrepreneurs, who sell scientific expertise and advice for a price. They may have come to their present role through university, industry, or government research. They are fundamentally salesmen, since much of their time and energy must be spent in acquisition of new contracts. New consulting groups are formed every year, but the mortality rate is exceptionally high, usually because of low initial financing, inadequate business procedures, or inability to acquire research contracts. With new firms, the successful completion of an initial contract is a major step toward survival, if not well-being.

Contract performance must remain consistently high, which means that a core of competent research people must be recruited and retained—an extremely difficult task in normal times, but an easier one in periods of restricted employment possibilities.

A basic, highly limiting fact of life for consultants is that in areas of controversy *if they do not develop data and reach conclusions favorable to their customers, they will very likely lose their clients.* This fact leads to some of the marginal practices discussed briefly in the chapter on "Ethics in Science" and to other scientifically unpalatable practices:

- Data are developed which will lead most directly to validation of the customer's position.
- Data are interpreted in such a way as to support the customer's position.
- Data are gleaned from public agencies for free, then packaged and sold to the customer at a high price.
- Data developed by consulting firms are often not published or otherwise publicly available to the larger scientific community.
- Consulting groups with a record of producing data and interpretations favorable to customers persist and prosper, while others disappear.

These and other ethical fringe practices have led to the application of charges of "scientific streetwalking," "selling your scientific soul to the devil," and "industry whore" to some industry consulting groups which too consistently find in favor of their customers.

Before concluding, it should be pointed out that not all, or even most, industry consulting groups are dishonest or unethical. Many attempt to provide scientifically supportable conclusions and recommendations on a variety of noncontroversial issues which lead to industry decisions. Many provide much-needed and balanced scientific perspective on complex, long-range problems, and do it well. All these activities constitute legitimate and valuable uses of scientific data and expertise.

A special breed of university scientists is also active in industry consulting. This group consists of faculty members with some credibility in their field and an acquired ability to assemble, synthesize, and analyze data relevant to a customer's needs. Universities are highly variable in their attitudes toward these extramural involvements; some permit substantial amounts, and others none. The more astute and experienced members of this consultant cadre are also called upon as expert witnesses in hearings and court cases. Some appear regularly in hearings conducted by agencies such as the Corps of Engineers, the Environmental Protection Agency, and the Food and Drug Administration. Others appear regularly as expert witnesses for utility companies, petroleum companies, or forest products companies. Often the thrusts of their testimony and conclusions are predictable, based on past performances and the best interests of their current employers.

An associated phenomenon which I have seen reenacted often is the propagation by the consultant/professor of graduate students who, after suitable apprenticeships, perform the same kinds of industry or agency support functions as their major professor. This is an entirely logical evolution, and the financial returns are attractive. These activities usually don't interfere with Ph.D. programs—in fact some thesis topics may be outgrowths of consultant projects. A small, worrisome element here has to be the totally pragmatic approach to research and to science that is instilled during this kind of maturation process.

THE SCIENTIST AS AN INDUSTRY SPOKESPERSON

An important area of science-industry interaction is in the courtroom or hearing room, as discussed briefly in the section on lawyers in the preceding chapter. Here the best game players from the industry research laboratory, public relations department, industry consultant group, or all of these confront the best scientific spokespersons from the federal agencies, various citizen action groups, or private foundations. Industry representatives usually take a low-key approach in issues relating to environmental degradation, trying to minimize the problems. The environmental action groups and the regulatory agencies, on the other hand, want to get as much factual information into the record and before the public as possible.

Polluting industries especially are willing to spend substantial sums to delay, impede, or prevent actions which will force them to change processes or carry out cleanup campaigns. Part of the expenditure of funds is to buy scientific expertise that can be counted on to support the industry position, which is always to minimize or deny the extent of demonstrable damage. Science, in these instances, is being used as a front—a camouflage—for environmentally destructive practices of the corporation. The degree to which some industries will invest in "safe" scientific advice (usually from consultants) often comes as a distinct shock to scientists unfamiliar with these procedures.

In some court cases involving actions against large industries (such as tobacco, mining, toxic chemicals, metals, and pharmaceuticals) a special cadre of industry spokespersons, who are not scientists but who have combined science with legal expertise, will act for the company. These people are usually part of company administration rather than the research laboratory and are skilled in attempting to suppress damaging information, in presenting slanted interpretations of data to support the company's position, and in trying to discredit expert witnesses called by the opposition.

Normally, the major contribution of expertise assembled against environmentally damaging industries comes from federal research laboratories. Competent scientists present arrays of data gathered in long-term monitoring programs—data which usually cannot be discounted by legal manipulations of industry lawyers, provided that the government scientists are also good game players and have done their homework.

SUMMARY

It may be that scientists as a group disagree with the concept of financial gain as a sole reason for existence or with total pragmatism as an operating philosophy. But whatever the reasons, relationships of most scientists with industry must be described as "uneasy." Many industries depend on a flow of new processes and products, some of which emerge from research laboratories. To this extent, industries are dependent on the research community, whether it is part of the company structure or whether it is bought with contracts given to universities or consulting firms.

Depending (as usual) on the individual, industrial research can be a wholly satisfying or a nightmarish occupation. Some scientists can prosper in a hard-driving, deadline-oriented, mission-satisfying atmosphere, while others are totally repelled. Some scientists find great satisfaction in applied research with its immediate payoffs, while others prefer long-term basic studies. Those considering jobs in industrial research need to evaluate their own preferences and make career decisions carefully. For the well-prepared, aggressive, and people-oriented scientist, industrial research can provide quick access to good salaries and management positions. The risks are commensurate with the advantages, however.

The new century promises to bring science-driven major changes in human existence—changes that are already beginning to emerge under the broad umbrella of "biotechnology." Genetic engineering will be the core activity for advances in medicine, pharmaceuticals, and food production. Those advances will undoubtedly shift the overall emphasis of industrial research.

CONCLUSIONS

Science consists fundamentally of "a search for truth and understanding through the acquisition and interpretation of factual information derived from observation or experimentation," and scientific research consists of "explorations directed toward understanding natural phenomena." Prime requirements for a life in science are adequate preparation, hard work, brilliant insights, intelligent analyses, and timely syntheses. If this were the totality, much of science could be characterized as satisfying but also a suburb of Dullsville. Fortunately there is another entire dimension to the practice of science, floating in free-form above the laboratory benches, the meeting rooms, and the conference mixers—providing pleasure and challenge to many. This I have described as "Scientific Game Playing"; it is really an aggregate of interpersonal strategies which ensure maximum rewards and satisfactions from a scientific career. It has a vital base in productivity and credibility in science but moves far beyond those essential but somewhat mundane ingredients.

This book is intended to be a training manual for scientific game players. Structurally, approaches to game playing have been described at three levels—basics, higher orders, and special cases. Looking backward briefly, I first dragged you through the nitty-gritties of game playing—approaches to writing and presenting scientific papers, attending meetings, chairing and organizing meetings, and even participating in the inevitable committee meetings. Only then did I feel you were prepared for such topics as promotions, acquisition of power, job changes, and ethics. After surviving all these chapters, I still felt there were special areas that deserved attention from game players. Of the many which might have been considered, several seemed important, including women in science and interactions with bureaucrats, news media, lawyers, politicians, the public, and industry.

Part One—the fundamentals—tends to be somewhat pedestrian, yet it is knowledge and use of fundamentals that win games, whether the playing field is a football stadium or the annual meeting of a scientific society. I consider such topics as writing and presenting scientific papers, chairing scientific sessions, and organizing scientific meetings as fundamentals—as significant and challenging learning experiences for the emerging game player. Until such fundamentals are mastered, the higher levels of scientific games are not easily accessible.

Part Two—higher orders of game playing—attempts to come to grips with less tangible strategic subjects, such as promotions, job changes, power, and ethics. The footing gets a little slippery in these areas, and there is less room for dogma and absolutes, but this is where the real action occurs.

Part Three—special cases—turns out to be a hodgepodge of special topics beginning with women in science, then moving briskly to examine relations with bureaucrats, news media, lawyers, politicians, the public, and industry. Each of these subject matter areas contains new opportunities for the strategist.

From the perspective of the closing pages of this book I remain convinced of my central and original thesis—that good science is great in itself, but is so much more rewarding, and often more fun, if scientists pay attention to the games being played and become active, informed participants in those games.

A good scientist begins with a better than average core of energy, intelligence, and perception. These are closely interactive with personal characteristics of productivity, insights (conceptual thinking), synthetic ability, analytic ability, enthusiasm, effective oral expression, and effective written expression. The good scientist can then be abstractly visualized as an interlocking matrix of individual traits, some inherent and some learned.

The *successful* and *complete* scientist superimposes on this core of excellence a thick frosting of interpersonal strategies based on some of the concepts in this book. Here, at long last, is the final synthesis.

So with this the game is over. I have taken you by the hand and led you across most of the major playing fields of science. You may have noticed that *virtually nothing has been said about scientific research itself*. This was deliberate, since the book is not about scientific research. It began with the assumption that the game player does excellent research, since this is the foundation of credibility on which all games must be based, but I have attempted to show that a whole universe of profitable and enjoyable games exists outside the laboratory or classroom door.

I close this discussion as I began it—sturdily supportive of the necessity for description and codification of the games played in science—and I look forward to confirmations, critiques, or nasty comments, as you see fit.

REFERENCES

Berne, E. 1976. *Beyond games and scripts*. New York: Ballantine Books.

Blau, P. 1973. *The organization of academic work*. New York: J. Wiley.

Boyd, R. S. 1998. Reining in the overload of information. *Miami Herald*, February 11, 1998, p. 17.

Broad, W. J. 1980. Would-be academician pirates papers. *Science* 208: 1438–40.

Broad, W. J. 1981. The publishing game: getting more for less. *Science* 211: 1137–39.

Carswell, D. 1950. Cabbages and kings. *Harvard Crimson*, June 14, 1950.

Clemmet, M. 1990. Toughest federal science jobs elude women. *The Scientist* 4 (20): 8.

Cole, J. R. 1979. *Fair science: women in the scientific community*. New York: Free Press.

Cole, J. R. 1981. Women in science. *American Scientist* 69: 385–91.

Cole, J. R., and B. Singer. 1991. A theory of limited differences: explaining the productivity puzzle in science. pp. 277–310 In Zuckerman, H., J. R. Cole, and J. T. Bruer (eds.). *The outer circle: women in the scientific community.* New York: W. W. Norton.

Cole, J. R. and H. Zuckerman. 1984. The productivity puzzle: persistence and change in patterns of publication of men and women scientists. pp 217–58 In Steinkamp, M. and M. L. Maehr (eds.) *Advances in motivation and achievement,* volume 2. Greenwich, CT: JAI Press.

Dean, C. 1998. After a struggle, women win a place "on the ice." *New York Times,* November 10, 1998, pp. 1, 4 (Science News).

Fox, M. F. 1989. Women and higher education: gender differences in the status of students and scholars. pp. 188–204 In Freeman, J. (ed.) *Women: a feminist perspective.* Palo Alto, CA: Mayfield Publications.

Fox, M. F. 1991. Gender, environmental milieu and productivity in science. pp. 221–43 In Zuckerman, H., J. R. Cole, and J. T. Bruer (Eds.). *The outer circle: women in the scientific community.* New York: W. W. Norton.

Goffman, E. 1972. *Relations in public.* New York: Basic Books.

Golde, R. A. 1981. Power comes out of the closet. *Harvard Magazine,* March-April, 1981, pp. 79–81.

Hafner, K. 1998. Physics on the Web is putting science journals on the line. *New York Times,* April 21, 1998, p. 3 (Science News).

Hayes, B. 1994. The World Wide Web. *American Scientist* 82: 416–20.

Heclo, H. 1977. *A government of strangers.* Cambridge, MA: Harvard University Press.

Hunt, M. 1981. A fraud that shook the world of science. *New York Times Magazine,* November 1, 1981.

Koeppel, D. 1998. Easy degrees proliferate on the Web. *New York Times,* August 2, 1998, pp. 17–19 (Education).

Kotter, J. 1979. *Power in management.* New York: American Management Institute.

Merton, R. K. 1957. Priorities in scientific discovery: a chapter in the sociology of science. *American Sociological Review* 22 (6): 635–59.

Merton, R. K. 1961. Singletons and multiples in scientific discovery: a chapter in the sociology of science. *Proceedings of the American Philosophical Society* 105 (5): 470–86.

Merton, R. K. 1963. Resistance to the systematic study of multiple discoveries in science. *Archives of European Sociology* 4: 237–82.

Merton, R. K. 1969. Behavior patterns of scientists. *American Scientist* 57 (1): 1–23.

Pelz, D., and F. Andrews. 1976. *Scientists in organizations: productive climates for research and development.* Ann Arbor, MI: Institute for Social Research.

Shenker, I. 1972. Ex-Pentagon aide lauds the bureaucrat-politician. *New York Times,* March 5, 1972, p. 31.

Sindermann, C. J., and T. K. Sawyer. 1997. *The scientist as consultant.* New York: Plenum Press.

Stockton, W. 1980. On the brink of altering life. *New York Times Magazine,* February 17, 1980.

Sullivan, E., W. Stewart, and H. Spalle. 1996. *External degrees in the information age.* Washington, DC: American Council on Education.

Taubes, G. 1993. Publication by electronic mail takes physics by storm. *Science* 259: 1246–48.

Valian, V. 1998 *Why so slow: the advancement of women.* Cambridge, MA: MIT Press.

Walker, T. J. 1998. Free Internet access to traditional journals. *American Scientist* 86: 463–71.

White, B. 1998. The World Wide Web and high energy physics. *Physics Today,* November, 1998, pp. 30–36.

Wilson, E. O. 1998. Scientists, scholars, knaves and fools. *American Scientist* 86: 6–7.

Yentsch, C. M., and C. J. Sindermann. 1992. *The Woman Scientist.* New York: Plenum Press.

Zuckerman, H. 1991. The careers of men and women scientists: a review of current research. pp. 27–56 In Zuckerman, H., J. R. Cole, and J. T. Bruer (eds.) *The outer circle: women in the scientific community.* New York: W. W. Norton.

Zuckerman, H., J. R. Cole, and J. T. Bruer (eds.) 1991. *The outer circle: women in the scientific community.* New York: W. W. Norton.

INDEX

Academia, 146, 187, 194, 214, 246, 271
  administrative bureaucracies, 240–241
  consulting and, 9–10
Acknowledgements, 33–34
Administration, 35
  managerial types, 154–155
  transitions to, 146–148, 151–154
  See also Science management
Administrative bureaucracies, 240–241
Advanced degrees, 194
Advisors, 83–84
Advisory committees, 133–135
Aggressiveness, 145, 146, 168
Annual meetings, 106–113
  chairperson, 108–109
  commercial firms and, 111–113
  format, 109–111
Assistants, 183
Attending scientific meetings
  funding attendance, 78–79
  international meetings, 81–86
  interpersonal relationships, 72, 74–75

organizing committee, 74, 76–77
presentation of papers, 72–73
residues of, 86–87
social activities and, 72, 75–76, 127
spouses at, 76
strategies for participation, 71–78
Audience, 63–66, 92, 96–98
Authorship, 15, 28–40
  acknowledgements, 33–34
  books, 38–39
  citations, 31–32
  comments on others' work, 32–33
  debate, 29–31
  editing, 18, 39
  in expanding research areas, 34–35
  first authorship, 28, 29–31
  formula for determining, 30–31
  priorities of information, 34
  in publish or perish climate, 35, 39
  rejection of manuscript, 37–38
  reviews, 35–37

sequence of authors, 29
support groups, 39–40
See also Editors; Journals;
    Publication

Babbage, Charles, 201
Banquet or luncheon speakers,
    101–103, 107–108
Biotechnology, 4, 20–23, 273
Books, 38–39
Broad, W. J., 49
Bureaucracy, 225–226
    administrative, 240–241
    government bureaucracies,
        239–240
    hierarchy, 237–238
    kinds, 239–247
Bureaucrats
    career bureaucrat, 226, 227,
        230–232
    characteristics, 234–237
    government bureaucrats,
        234–236
    politician-bureaucrat, 226,
        227–230
    time-server bureaucrat, 226,
        232–234

Career
    first decade, 155–156
    first job, 163–164
    gopher syndrome, 164–165,
        263
    in industry, 263
    life stages, 150–155
    mature scientist, 157–158
    middle years, 156–157
    promotion, 145–146
    as satisfying, 2–3
    See also Academia; Government;
        Industry; Transitions
Career bureaucrat, 226, 227,
    230–232
Carswell, D., 48
Castrations, scientific, 202–204

Chairperson, 104
    chairperson's creed, 126–127
    characteristics, 161–162
    for committee meetings,
        125–127, 129–131, 140
    games played by, 130–131
    international meetings, 119, 120
    selection of, 108–109
    session follow-up, 98–99
    special assignments, 100–103
    See also Scientific sessions
Citations, 31–32
Civility, 196
Clemmet, M., 221
Coffee breaks, 95, 127, 245
Cole, J.R., 6, 7, 215, 216
Collegiality, 8, 216–217
Commerce Department, 180–181
Commercial organizers, 111–113
Committee meetings
    advisory committees, 133–135
    chairperson, 125–127, 129–131,
        140
    computer conferences, 137–140
    games, 130–132
    members' guide, 128–130
    paper products (reports),
        135–137
    secretary, 136
    value of, 128–129
    women and, 132–133
Communication, 16, 106–107, 122
Computer conferences, 137–140
Consulting, 4, 9–12
    ethics and, 194–195
    guiding principles, 11–12
    in industry, 269–271
Convenor, 115
Coordinators, 237
Corporations, 265–266, 268
Credibility, 16–17, 159, 177,
    204–205, 253, 277

Data, 199–202, 270, 272
Dean, Cornelia, 212

Directors, 177–178
Disclosure process, 3
Discussants, 93, 96, 97, 109–110
Discussions, 67–69, 73–74, 114
    audience participation, 96–98

Eastern Europe, 118
Ectoparasitism, scientific, 201
Editors, 36–37, 45–49, 179, 182
    workshops and, 116–117
    See also Journals
Electronic communication. *See*
    Internet; Video conferencing
Elitism, 14–15
E-mail, 17, 19–20
Energy, 184, 277
Enthusiasm, 55, 62, 145, 159
Entrepreneurial ventures, 266–268
Entrepreneurs, 4, 9, 269
Environment, 257–258, 272
Esprit de corps, 15
Ethics, 11, 21, 141–142, 191–192,
    206
    consulting and, 194–195
    deflating the instant scientist,
        204–205
    ethical activities, 197–199
    industry and, 270, 272
    questionable activities, 192–197
    scientific castrations, 202–204
    use and abuse of data, 199–202
Executive committee, 122
Executive groups, 131
Executive secretary, 111
Executive summary, 137
Expert witnesses, 253–255, 271
Extrapolating, 200

Family issues, 156, 166
Fast track, 146–149
Federal scientific career
    bureaucracy, 232, 233
Feminist movement, 210, 223
First authorship, 28, 29–31
Follow-through, 115–116

Fraud, 204–205
Fudging, 201
Funding, 14–15, 78–79, 180–181,
    256
    bureaucracies and, 235, 241–247
    review processes, 243–247
Funding agencies, 151–152

Game playing, 1–3, 32, 130–132,
    263–264, 275–277
Gelb, Leslie, 227
Genetic engineering, 21–22, 273
Gentle manipulator, 162
Glass ceiling, 219–220
Golde, R. A., 186
Gopher syndrome, 164–165, 263
Governmental laboratories,
    187–188
Government bureaucrats, 234–236
Government of Strangers, A (Heclo),
    228
Government science management
    bureaucracies, 239–240
Grants. *See* Funding
Grantsmanship, 242–243

Hayes, B., 19
Heclo, Hugh, 228
Honoraria, 93
Humor, 102–103

Impostor Syndrome, 222
Industry, 10, 261–262, 273
    administrative bureaucracies,
        240–241
    consulting in, 269–271
    contracts, 262, 269–270
    ethics and, 270, 272
    game rules, 263–264
    kinds of research laboratories,
        264–268
    scientist as spokesperson,
        271–272
Infighting, 34–35, 195–196
Informal paper presenter, 73

In-groups, 80, 139, 144, 179, 184
Instant scientist, 204–205
Institutes, 187
Intellectual property, 15, 19,
    199–202
Intelligence, 184, 277
International Council for the
    Exploration of the Sea, 85
International meetings, 81–86
    chairperson, 119, 120
    Eastern Europe, 118
    executive committee, 122
    nationalism, 122
    organizing, 117–123
    political statements, 87
    representatives of countries, 118,
        122
    review papers, 118–119
    scientific advisors, 83–84
    speakers, 120–121
    translation services, 120, 121
Internet, 4, 17–20, 22
    guidelines, 18–20
Interpersonal relationships, 3, 12,
    25–26, 144, 275
    nonscientific positions and,
        148–149
    with politicians, 256–257
    science management and, 15–16
    scientific meetings and, 72,
        74–75
    women and, 7–8
    at workshops, 114, 116
Interpretations, 254, 257
Interruptions, 131
Introductions, 56, 93, 110

Jargon, 22
Journals, 32–33, 203
    editors, 45–49, 182
    format, 46
    rejection of manuscript, 37–38
    reviewers and, 40–45
    reviews, 35–37
    See also Editor

Journeyman scientist, 151
Junior scientist, 150–151, 155–156,
    193–194, 203, 269

Keynote speech, 100–101
Kotter, J., 176–177

Lawyers, 253–255
Lay people, 259–260
    lawyers, 253–255
    news media, 249, 250–253
    politicians, 235, 255–257
Least Publishable Unit (LPU), 49–50
Libraries, 17–18
Life stages of scientist, 150–155
Limited differences, 216
LPU. See Least Publishable Unit

Management training, 153
Managerial types, 154–155
Manufacturing data, 201
Manuscript
    proofreading, 46–47
    rejection of, 37–38
    writing style, 45–48
    See also Authorship; Publication
Massaging, 200
Maximum Publishable Unit
    (MPU), 50–51
Maximum Recycling Quotient
    (MRQ), 50–51
Media. See News media
Mediocrity, 165–166, 187–188
Meetings. See Annual meetings;
    Attending scientific meetings;
    Scientific sessions
Mentoring, women and, 8,
    216–217
Merton, R. K., 202
Microphones, 95
Mini-symposia, 109
Minutes, 131
Mirror writing, 201
MPU. See Maximum Publishable
    Unit

MRQ. *See* Maximum Recycling Quotient

National Academy of Sciences, 7
National Education Association (NEA), 213
National Oceanic and Atmospheric Administration (NOAA), 137–138
NEA. *See* National Education Association
News media, 249, 250–253
   misinformation, 252–253
   television personalities, 258–259
NOAA. *See* National Oceanic and Atmospheric Administration
Nobel Prize, 7
Nonverbal ploys, 131–132
Novice scientist, 150–151

Office of Management and Budget (OMB), 234–235
OMB. *See* Office of Management and Budget
Open organization, 15
Optimum Publishable Unit (OPU), 50–51
OPU. *See* Optimum Publishable Unit
Organizing scientific meetings, 74, 76–77, 105–106
   annual meetings, 106–113
   commercial organizers, 111–113
   financial considerations, 112–113, 119–121
   international symposia, 117–123
   program committee, 111
Overview papers, 110, 116

Panels, 110
Papers, 93, 96, 99, 104, 109
   committee meetings and, 135–137
   international meetings and, 118–119
   overview papers, 110, 116
   paper flood, 49–52
   scientific meetings and, 72–73
   See also Presentations
Participants, 91–92, 104, 114
   committee members, 128–130
   See also Attending scientific meetings
Patents, 22, 263
Perception, 184, 185–186, 277
Plagiarism, 201
Policymakers, 237–238
Political skills, 145–146, 185–186
Politician-bureaucrat, 226, 227–230
Politicians, 235, 255–257
Politicization, 112, 134–135
Power, 16, 141
   in business organizations, 176–177
   chairperson and, 180
   defined, 175–176
   of directors, 177–178
   editing and, 179, 182
   external signs of, 182
   funding and, 180–181
   "great one," 178–179
   in-groups and, 179
   kinds, 178–182
   organizational variations, 187–189
   politician-bureaucrat and, 227–230
   power figures, 183–186
   research groups and, 179–181
   of scientific peers, 182
   strategies, 183–186
   women and, 218–222
Premature disclosure, 201
Presentations, 53
   audiences, 63–66
   catastrophes, 60
   computerized projection, 60–62
   discussions, 67–69, 73–74

enjoyment of, 64–67, 90
enthusiasm, 55, 62
guidelines, 56–64
how not to present, 54–55
humor, 65–66
introduction, 56
lighting, 55, 59–60
speaking skills, 62–63
structure, 56
timing of papers, 96
visuals, 54, 55, 56–62, 92–93
Priorities, 34, 202
Producers, 237
Product development, 10–11
Professionalism, 154–155
Professional scientist, 158–163
   chairperson, 161–162
   characteristics, 158–160
   gentle manipulator, 162
   mature scientist, 157–158
   survivors, 160–161
   unprofessional professionals,
      42–44
   See also Career; Transitions
Professional team leader, 151
Projectionist, 94–95
Promotion guidelines, 145–146
Proprietary information, 22, 198
Public, 257–259
Publication, 10, 145
   electronic publication, 17–19
   infighting, 34–35
   methodology, 27–28
   paper flood, 49–52
   ploys, 48–49
   reviewers, 40–45
   rewrites, 35, 46
   by women, 5, 6–7, 215–216, 223
   See also Authorship; Editors;
      Journals; Manuscript

Rapporteur, 136–137
Registration, 93, 121
Reports of committee meetings,
   135–137

Reputation, 193, 203, 253, 254
Research laboratories, 22–23
   entrepreneurial ventures,
      266–268
   industrial research laboratory,
      188, 264–265
   new acquisition of diversified
      corporation, 265–266
   power and, 179–181
   See also Team research
Research positions, 169, 170
   See also Team research
Retirement, 171–172
Review committees, 203
Review papers, 118–119
Review processes for funding,
   243–247
Reviews, 18, 33, 35–37, 46, 179
   guidelines, 41–42
   role of reviewers, 40–45
Reward Index (RI), 51
Rewrites, 35, 46
RI. See Reward Index
Role-playing, 128

Salary, 213–214, 223
Satellite people, 184
Sawyer, T.K., 9
Science, 32, 49
Science management, 4, 151
   funding, 14–15
   government bureaucracies,
      239–240
   managing for elitism, 14–15
   team research and, 13–17
   transition to, 146–148
   women and, 8, 218–222
   See also Administration
Scientific advisors, 83–84
Scientific castrations, 202–204
Scientific sessions, 89–90
   chairperson, 90
   coffee breaks, 95, 127
   discussions, 93, 96–98
   early planning, 91–93

introductions, 93, 96
keynote speech, 100–101
meeting summarizer, 103
papers, 93, 96, 99, 104
registration, 93, 121
selection of participants, 104
session follow-up, 98–99
See also Chairperson
Scientific societies, 7, 10, 12
annual meetings, 106–113
election to, 79–81
Scientific street-walking, 194–195,
253, 270
Sea Grant Program (Department of
Commerce), 180–181
Self-exposure, 3
Senior scientist, 152–153, 158
Series publications, 47
Sindermann, Carl J., 9, 213
Singer, B., 216
Site-visitation committee, 244–247
Slanting, 201
Small-group conferences. *See*
Workshops
Smoothing, 200
Social activities, 72, 75–76, 127
banquets and luncheons,
101–103, 107–108
coffee breaks, 95, 127, 245
Speakers, 10, 96, 104
banquet or luncheon speakers,
101–103, 107–108
at international meetings,
120–121
keynote speaker, 100–101
meeting summarizer, 103
Speaking skills, 62–63
Special assignments, 100–103
Special interest areas, 207–208
See also Bureaucracy; Industry;
Lay people
Specialty areas, 10–11, 17, 34, 179
Stockton, W., 34–35
Strategy, 141–142
losing strategies, 163–166

power and, 183–186
See also Ethics; Power;
Transitions
Street-walking, scientific, 194–195,
253, 270
Success, 143–145, 277
Supervisors, 14, 188–189
Supervisory skills, 153
Survivors, 160–161
Sycophants, 184

Team leader, 151
Team research, 4, 13–17, 262
See also Industry; Research
laboratories; Science
management
Technicians, 31
Television personalities, 258–259
Territorial limits, 34
Three option ploy, 235
Time, 20
Time-server bureaucrat, 226,
232–234
Transitions, 141, 143–146, 236
to administration, 146–148,
151–154
eras in professional career,
155–158
fast track, 146–149
guidelines, 152–153
life stages of scientist, 150–155
losing strategies, 163–166
moving on, 166–170
moving out, 170–172
promotion, 145–146
retirement, 171–172
See also Career
Translation services, 120, 121
Traverse, Joan, 211
Treasurers, 112
Turf establishment, 131–132

Understudies, 183
United Nations, 85, 117
Universities. *See* Academia

Unprofessional professionals,
    42–44

Valian, Virginia, 209
Verbal ploys, 131–132
Video conferencing, 138–139
Visuals, 92–95

Whistle blowers, 246–247, 263–264
White, B., 18
WHO. *See* World Health
    Organization
"Why So Slow: The Advancement
    of Women" (Valian), 209
Women, 4, 5–8, 209–210
    abandonment of science, 5, 6,
        222
    children and, 8, 217
    committee meetings and,
        132–133
    current status, 210–215
    gender gap in management,
        220–222
    impediments to equality,
        215–218
    mentoring and, 8, 216–217
    numbers of in science, 212–213

perceptions of by male
    colleagues, 217–218, 220
position, 214
publication and, 5, 6–7, 215–216,
    223
recognition, 5–6, 7, 214–215, 223
salary, 213–214, 223
science management and, 8,
    218–222
Workshops, 86, 113–117
    convenor, 115
    discussion leaders, 114
    editors and, 116–117
    follow-through, 115–116
    format, 114–115
    interpersonal relationships at,
        114, 116
    length, 114
    location, 115
    participants, 114
World-class scientist, 85–86
World Health Organization
    (WHO), 84–85

Yentsch. C.M., 213

Zuckerman, H., 6, 7, 215